Wildflowers, Trees, *and* Shrubs ...Texas

D1293470

Wildflowers, Trees, and Shrubs of Texas

REVISED EDITION

DELENA TULL
GEORGE OXFORD MILLER

TAYLOR TRADE PUBLISHING
A Lone Star Book
Lanham • Boulder • New York

Photography by George Oxford Miller
Illustrations by Margaret Campbell

Published by Taylor Trade Publishing
A Member of the Rowman & Littlefield Publishing Group
4501 Forbes Boulevard, Ste 200
Lanham, MD 20706
Distributed by National Book Network

Library of Congress Control Number: 2002 116372

ISBN 1-58907-007-0 (alk. paper)

♾™ The paper used in this publication meets the minimum
requirements of American National Standard for Information
Sciences–Permanence of Paper for Printed Library Materials,
ANSI/NISO Z39.48–1992.
Manufactured in the United States of America.

Contents

QUICK GUIDE TO SPECIAL KEYS (BY GENUS)

Preface

Say: Nature in its essence is the embodiment of My Name, the Maker, the Creator. . . . Nature is God's will and is its expression in and through the contingent world.

—*Bahá'u'lláh*

Plants whisper the secrets of the universe if we listen. They reveal the majesty and mystery and wonder and wisdom inherent in all life. Not just bizarre plants from exotic places, but the wildflowers, trees, shrubs, cacti, vines, mosses, and lichens in our own backyards.

We hope this book helps aquaint you with some of our closest, though often overlooked, neighbors. Learning the names of the plants, as with people, helps us appreciate their unique characteristics. Knitting close bonds with the life forms around us, like making new friends, enriches our lives. As we understand and value nature more, we come closer to the creative force that permeates this beautiful, ever-changing planet, and which resides within each of us.

The passage from Bahá'í scriptures is reprinted from *Tablets from Bahá'u'lláh* © 1978 by the Universal House of Justice with permission of the publisher. Bahá'í World Centre Publications, Haifa, Israel.

Many thanks to the members of the Native Plant Society of Texas who have assisted us and encouraged us over the years, in particular Carroll Abbott, Marshall Johnston, Benny Simpson, Billie Turner, Marshall Enquist, and Landon Lockett (for his up-date on the Texas native palms). Though some of our friends are no longer with us in the physical realm, we know they lovingly watch over us all.

VEGETATIONAL AREAS OF TEXAS

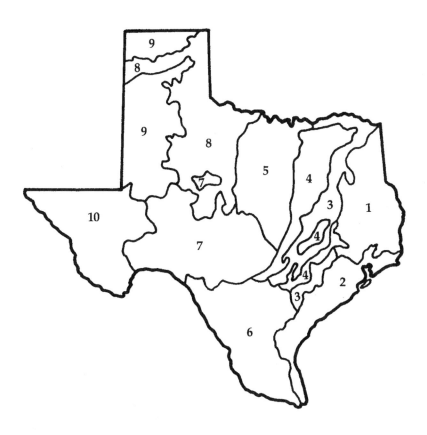

1	Pineywoods	6	South Texas Plains
2	Gulf Prairies and Marshes	7	Edwards Plateau
3	Post Oak Savanna	8	Rolling Plains
4	Blackland Prairies	9	High Plains
5	Cross Timbers and Prairies	10	Trans-Pecos, Mountains and Basins

© The Nature Conservancy Magazine, Vol. 32, No. 5.

1

How to Use This Book

Welcome to a field guide to the native and naturalized plants of Texas. This book includes many of the wildflowers, trees, bushes, vines, and weeds you'll come upon as you visit the outdoors in this vast state. In addition, you'll find a number of cacti and other succulents, as well as an assortment of peculiar looking things that grow in and around trees and rocks. We have developed a number of plant keys to assist you in distinguishing similar looking plants from each other. And we have grouped the wildflowers by color, to make it easier to locate the photograph of the species you find.

More than 650 species of plants are included in the book, with nearly 400 color photographs. We selected those plants that you would encounter most often in the various regions of the state, plus some uncommon species that are just too beautiful to pass up.

Numerous regional field guides have been published in the last few years. With the great diversity of plants in the state, a guide to your region would be a valuable supplement to this book. Throughout the book we have included brief notes regarding edible, poisonous, and useful qualities of the plants, including their uses as dyes for fibers. If you are interested in finding out more about wild edible and useful plants in Texas, refer to Delena Tull's book *Edible and Useful Plants of Texas and the Southwest* (University of Texas Press) which includes recipes; information on teas, spices, and medicinal herbs; instructions for dyeing wool with natural dyes, and for preparing fibers for weaving and papermaking; and valuable information on poisonous plants. For information on low-maintenance landscaping and planting for year round color, refer to George Miller's book *Landscaping with Native Plants of Texas and the Southwest* (Voyager Press). For a list of additional books on native plants, see the reference list in the back of the book.

WHAT IS A WILDFLOWER?

For many Texans, the term *wildflower* calls to mind fields of bluebonnets or Indian blankets. Technically, the term refers to more than just small herbaceous plants. Trees, shrubs, vines, cacti, and yuccas all grow wild and produce flowers; thus, all can be considered wildflowers. In this book, we have chosen to limit the term *wildflower* to the herbaceous annuals and perennials, those fairly small flowering plants that bloom annually, produce seeds for next year's crop, and then die back completely or partially in the heat and drought of summer or cold of winter.

For ease in identification, we have separated the photographs and plant descriptions into five groups: herbaceous wildflowers; trees and shrubs; cacti, yuccas, agaves, and other succulents; vines; and miscellaneous plants that don't seem to fit anywhere else. Within each section, the species are arranged in alphabetical order by the scientific family name and then by genus.

The herbaceous wildflowers are grouped by color (red, pink, blue, and purple; yellow; white and green). Though it may seem odd to group red and blue flowers together, many species naturally range in color from various shades of red to blue. This trait is due to the presence of anthocyanins, pigments that produce either blue or red colors. Other red flowers are colored by carotenoids, which also produce yellow and orange colors.

Wildflowers may be annuals, biennials, or perennials. *Annuals* emerge, bloom, produce seeds, and die back to the ground in one season or year. The plants will not return the next year unless they have produced seeds. *Biennials* may produce only leaves the first year, often in a basal rosette; they will produce flowers and seeds the second year and then die. *Perennials* often produce foliage, flowers, and seeds the first year, but the rootstock and sometimes part of the aboveground stem survive extremes of drought and temperature, and new foliage and flowers emerge year after year.

The term *herbaceous* refers to plants that die back each year. It also usually refers to a plant that is not woody. However, a number of perennial wildflowers have a woody base or stem that does not die, although most of the aboveground part of the plant is herbaceous.

In this book, *herbaceous vines* are those that die back in winter, and *woody vines*, although they may lose their leaves, are those that maintain an aboveground vine year-round. *Trees* and *shrubs* are woody plants that do not die back every year. (With the trees and shrubs, however, we have included a few plants that become large and bushy, such as rattlepod, even though they do die back in winter).

Many people use the term *cactus* to include anything with thorns, but many trees and shrubs bear thorns. Refer to the key to cacti, yuccas, agaves, and other succulents to learn the characteristics that distinguish these bizarre members of the plant kingdom.

If you are not sure which category includes your plant, then perhaps the following bit of folk wisdom will assist you. A friend of ours once said, "A tree is something that knocks you down if you run into it." Although a shrub or bush may not knock you down, it will certainly impede your progress. And the poor herbaceous wildflower is readily trampled underfoot. A vine might be underfoot or overhead, but it always seems to be crawling out of reach.

If you have a plant that doesn't fit in any of the above categories, glance through the miscellaneous section, which includes some things that you may know as weeds, and a few nonflowering plants, such as mosses. In addition, we have included lichens (which are composed of algae and fungi) and galls (which are growths on plants). Several large groups of flowering plants are totally ignored by this book for no better reason than that their flowers are so minute as to require a microscope for identification. Among them are the grasses, a diverse family with more than five hundred species in Texas.

HOW TO USE A PLANT KEY

The purpose of a key is to allow you to identify a plant through the process of elimination. Each step presents two possibilities, a or b. Choose the one that best describes the plant you want to identify. If you are unfamiliar with some of the terminology, the illustrations will help you make your decision. Also, you can refer to the glossary and the illustrated glossary at the back of this book.

Within the key, the numbers in parentheses indicate the previous step, in case you have to backtrack. The number at the end of each statement directs you to the next set of choices. As you proceed, step by step, you narrow the possibilities of what the plant might be. Finally, you will reach a step that identifies the plant by family, genus, or species and lists the page number in the text. Look at the photograph and text description to verify that you have correctly identified the plant.

Using a key requires accurate observation of features. Pay particular attention to whether the leaves are simple or compound, alternate or opposite. You may have to examine numerous leaves to determine the characteristics. Once you become familiar with the key, you can use the quick key to skip some of the preliminary steps in the longer key. For example, if you know the plant has simple leaves, you can advance immediately to step 25 and proceed from there. Using a key becomes easier with familiarity and practice.

PHOTOGRAPHS AND DESCRIPTIONS

Each photograph is accompanied by a description of the plant. For trees only, when an exact height is given, it refers to the state champion tree. In many cases, there are additional closely related species that are not pictured. Some of them might be confused with the species photographed, and others are quite distinctive. Notes at the end of the description give information about related and similar species, and where space allows, descriptive information on the related species is given. The comment *hand lens* means that the reader may need to use a magnifying lens to see the distinguishing features of the plant.

RANGE MAPS

Range maps accompany the descriptions of each species. The maps will enable you to form an immediate visual image of where the plant is known to grow. These maps are not complete. For many species, little documentation is available. We developed the maps from a variety of sources: personal field observations, correspondence with other botanists, visits to herbariums around the state, and information in botanical manuals, particularly *Manual of the Vascular Plants of Texas* by Donovan S. Correll and Marshall C. Johnston. Even with all these sources of data, the maps sometimes contain inaccuracies. The distribution of Texas plants is far from fully documented. The reasons include lack of funding for extensive field verification, the ongoing effect of habitat destruction, the introduction of foreign competing species, and the cultivation of native species outside their natural ranges.

When records were inadequate, we had to guess where the boundaries might fall. We have attempted to document the distribution of each species, including both the natural range and the areas where the plant has become naturalized (adapted to a new area). Data on species with recently expanded ranges, such as the bluebonnet, are poorly documented in herbariums.

The University of Texas (http://www.biosci.utexas.edu/prc/) and Texas A&M University (http://www.csdl.tamu.edu/FLORA/) both now have excellent web-sites with extensive information on Texas plants. The UT Flora of Texas database provides information on plant specimens that have been collected, including county of collection. If maintained, this will provide a wonderful resource to assist in identifying the ranges of species.

WHY USE BOTANICAL TERMINOLOGY?

All of us have names for some of the plants that grow around us. *Dandelion*, *oak*, and *maple* are names that many people recognize. Not so frequently heard are *Taraxacum*, *Quercus*, and *Acer*, the scientific names for these plants. Why bother to learn these botanical terms? Perhaps you will never have a need to use them. However, if you move to a different area of the state or the country, you may find that people use different common names from those you have learned. Even in one region, a single plant may have several common names. In addition, two different plants may share the same common name. For example, *pigweed* may refer to a member of either the genus *Amaranthus* or the genus *Chenopodium*.

International congresses decide on the rules that determine the validity of scientific names. The names are in Latin and are recognized by botanists worldwide. If you learned the scientific names of plants in Texas, you could communicate about them with a plant enthusiast from the other side of the world.

The scientific name of a plant consists of two parts: the generic name, which is first, and the specific epithet. For example, the scientific name of the Texas bluebonnet is *Lupinus texensis*. There are six species of bluebonnets in Texas, all in the genus *Lupinus*. When several species are listed for one genus, the genus name may be abbreviated, such as *L. texensis*. The specific epithet *texensis* distinguishes the Central Texas species seen growing in abundance along highways in the spring from the tall Big Bend bluebonnet, *Lupinus havardii*. The generic name *Lupinus* is necessary to distinguish this *texensis* from all other plants in the world with the same epithet (such as *Colubrina texensis*, a Central Texas shrub). Members of a species may have enough local variation to warrant the designation of several subspecies or varieties (for example, *Quercus pungens* var. *vaseyana* and *Garrya ovata* subsp. *lindheimeri*). Many generic names and specific epithets have as their root the name of a botanist, such as Drummond, Engelmann, or Maximilian. Often the botanist named is the one who first discovered the species. In some cases the classifying botanist names a plant in honor of another botanist.

Unfortunately for the amateur botanist, scientific names occasionally change as new information on plant identity is provided through field research. Old names that are no longer considered valid may still appear in botanical literature. In this book, we have included in parentheses next to the species name the synonym or alternate for the species name. A synonym is a name that was used formerly, or an alternate name is given in cases where there seems to be a disagreement about which name is more appropriate.

In some cases, botanists disagree as to which name is appropriate. Why is there such disagreement? To answer that question, we must first pursue the question, What is a species? In layman's terms, two types of plants or animals are considered tò be distinct species if they do not mate with each other. However, the layman's definition does not take into account the many gray areas, the many examples of cross-species hybrids, occurring in nature. For example, we all know that a dog and a coyote are different types of animals, and yet, under some circumstances, they will mate.

The species designation is the basic unit for classifying living things. Taxonomy is the science of classifying, or naming, things. Though the concept of the species is complex, taxonomists generally define a species as a population of individuals that share the same basic characteristics and are more or less genetically isolated.

Species that have certain genetic and structural similarities are classified as one genus. Genera (the plural for *genus*) with similar characteristics are classified as one family. Botanists also use larger categories, but these three are the most useful for field identification.

These groupings may sound straightforward and easy to define. However, someone has to decide which characters of a species make it unlike another species, and which characters shared by several species are important enough to unite them into one genus. In examining a botanical manual, one finds that some characteristics used to differentiate between two species are readily apparent (the number of petals on a flower, or the arrangement of leaves in pairs) whereas other characteristics are inconsistent or difficult to see (the length of sepals, or the type of hairs on a leaf).

Part of the problem is that nature doesn't always fit into the nice, neat compartments that we design for it. The categories selected by botanists are by necessity artificial.

Plants that seem dissimilar enough to be considered separate species may in fact hybridize when growing next to each other. Certain groups of plants (oaks, for example) seem particularly prone to forming hybrids. Closely related parent species often produce viable hybrids that can produce offspring or form back-hybrids with the parent species. A population of plants consisting of a variety of partial hybrids may exhibit a number of differing characteristics and is known as a hybrid swarm. Polyploidy, the remarkable ability of plants to form hybrids by doubling the chromosome numbers, enables many plants that are not as closely related to produce viable hybrids as well.

Under certain conditions, plants can produce new species rapidly when species that naturally occur in the same area hybridize (sympatric speciation). Hybrids commonly occur in the plant kingdom, and occasionally a hybrid population fills a niche slightly different from those of the two parent populations. If it maintains itself as a separate population, it can be considered a new species. (By the way, sympatric speciation, though common in flowering plants, is virtually unheard of in animals; the hybrid animal is usually sterile.)

In addition to speciation by hybrid swarms, plants also form new species by allopatric speciation (the speciation method common for animals). For example, if a physical barrier separates a population into two groups, over time they may form two distinct populations adapted to different climatic, soil, or moisture conditions. These two new populations, if brought back into contact with each other, may no longer be able to produce viable offspring and therefore would be considered separate species.

Frequently, variation in a population occurs along an environmental continuum. For example, plants of the same species may grow along a geographical continuum from Canada to Texas. Though plants in adjacent populations along the continuum are capable of reproducing with each other, the populations at the two extremes may not be able to reproduce if brought into contact with each other. At this point, one may well question the validity of the concept of a species. Perhaps it makes no sense to put names on things at all. On the other hand, categories help us catch a glimpse of the intricate order in the world of nature and help us understand something about the complex web of life that surrounds and sustains us.

As species of plants are examined, through extensive field work and examination of herbarium specimens, plant taxonomists, the scientists who select the names for plants, periodically review and may revise the species, genus, and sometimes even the family names for plants. The recent advent of DNA analysis provides a new tool to assist in ascertaining relationships between plants. In addition to the University of Texas and the Texas A&M University web-sites, two additional web-sites are extremely useful in keeping track of scientific name changes: the USDA plant web-site (http://plants.usda.gov/) and Texas A&M web-link to the Synonymized Checklist of Vascular Flora of the United States, Puerto

Rico, and the Virgin Islands (http://www.csdl.tamu.edu/FLORA/ b98/check98.htm) (I found that the string query was the easiest way to search the vascular flora database. This wonderful web-site links species names to range maps). We used all of these resources in up-dating the names for the latest edition of the book. ❧

2

Regions of Texas

With the Rocky Mountains to the west, the Great Plains to the north, the Chihuahuan Desert to the south, and the Gulf of Mexico to the east, Texas lies at the biological crossroads of North America. From east to west, the elevation varies from sea level to 8,700-foot mountains, the moisture from 60 inches to 8 inches annually, and the soils from rich clays to deep sand or limestone caliche. From north to south, the temperature varies from an average of 110 days below freezing annually in Amarillo to only 2 days in subtropical Brownsville.

Because of the wide range of physical conditions, most plants grow only in specific areas within the state. Texas has ten major vegetational regions (see map), each with a distinct association of plants. Though each zone is characterized by certain dominant trees, shrubs, grasses and wildflowers, a region is by no means a homogeneous blend of plants. A particular species may grow only in a limited area within its vegetational region, or it may spread across several regions. Human designations and divisions of nature are necessarily arbitrary and therefore inexact. To provide range information as accurately as possible in this book, we often describe a plant's distribution by referring to both its geographic range and its vegetational region. The geographical areas we reference are described below.

East Texas—a broad term designating everything east of a line connecting Houston, Austin, and Dallas. This flat to hilly region includes blackland prairies, post oak-hickory woodlands (the Post Oak Savannah), and dense pine-oak forests (the Pineywoods). It has a relatively high annual rainfall (35–60 inches), with sandy and sandy loam soils and rich calcareous clays. The Big Thicket, an area of dense pine and hardwood forests and swamps, covers Newton, Jasper, and Tyler counties, and portions of surrounding counties in Southeast Texas.

Coastal Texas—the flat, humid area between the coast and a line from Beaumont to Houston, Victoria, and Brownsville. Grassland prairies, swamps, and marshes make up the area. The wetlands support millions of migrating waterfowl in the winter.

South Texas—also known as the Rio Grande Plains, the flat to hilly area south of a line connecting Corpus Christi, San Antonio, and Del Rio. The area not adjacent to the coast experiences periodic drought. The western half may have as little as 16 inches of annual rainfall, with close to 30 inches in the east. The area around Brownsville, the Lower Rio Grande Valley, has a subtropical climate and rarely experiences sustained freezing temperatures.

West Texas—the region west of Austin and Dallas and north of San Antonio. West Texas includes Central Texas, North Central Texas, the Edwards Plateau, the Panhandle, and the Trans-Pecos.

Central Texas—roughly includes most of the Edwards Plateau, extending somewhat to the north.

North Central Texas—the region of rolling hills extending from Dallas, south to Waco and west to Wichita Falls. The area has black-land prairies with clay soils, interspersed with oak woodlands (the Cross Timbers) on sandy loam soils. Extreme summer and winter temperatures and moderate rainfall (25–40 inches) characterize the climate.

Edwards Plateau—a specific geographic region, also known as the Hill Country, 1,000 to 3,000 feet in elevation. It extends west of Austin roughly to the Pecos River, and from Del Rio north to around Abilene. Spring-fed rivers cut rugged canyons, which harbor rare plants, and oak-juniper-mesquite savannahs cover the rocky, limestone hills. The rainfall varies from moderate (30 inches) in the east to semiarid (15 inches) in the west. Summer temperatures can be extreme, with frequent drought, but winters are generally mild.

The Panhandle—an area with few large trees extending roughly north of San Angelo and west of Wichita Falls. It includes the High

Plains (or the Llano Estacado) and the Rolling Plains. The Caprock Escarpment separates the two and forms Palo Duro Canyon near Amarillo. Thousands of playas, or shallow lakes, dot the High Plains, a flat plateau 3,000 to 4,500 feet in elevation. The Rolling Plains are a southern extension of the Great Plains of the central United States. Soils in the Panhandle vary from clays to deep sands and sand dunes. *Quercus havardii*, a three-foot shin oak, covers thousands of acres of sand, creating perhaps the largest oak forest in North America. The Panhandle has extreme summer and winter temperatures and experiences frequent drought (15-30 inches annual precipitation).

Trans-Pecos—the area west of the Pecos River. It ranges from 8,700-foot mountain peaks to the arid basins and valleys of the Chihuahuan Desert. Soils include those derived from igneous and limestone rocks, saline flats, and deposits of deep sand. The vegetation varies from boreal forests of fir and ponderosa pine to cacti, grasses, and shrubs adapted to extreme drought. The area experiences extreme drought and heat in summer and mild or bitter temperatures in winter. The desert receives 8 to 12 inches of rainfall annually, and as much as 20 inches falls in the high mountains. ❧

3

Diversity in the Plant Kingdom

If you ever feel bored with life and you want to see something astonishing, pack a lunch and take a trip to your backyard. The diversity in the plant kingdom can supply a lifetime of wonder and amazement. As though following a cosmic command, plants have "multiplied and filled the earth." From the Arctic Circle to the hottest desert, from the highest mountain to the ocean depths, plants don't just survive—they thrive. No single strategy could enable plants to adapt to the diversity of growing conditions around the world. The complex and intricate strategies plants have evolved to absorb energy, combat harsh conditions, reproduce, and avoid being eaten are stranger than any science-fiction fantasies humans will ever dream up.

Fossil records indicate that terrestrial plants have covered the earth for about 400 million years. By producing their own food through the complex process of photosynthesis, plants provide the two ingredients necessary for the animal kingdom to survive: atmospheric oxygen and a primary source of food. Plants absorb carbon dioxide from the air and, using the energy in sunlight as well as water and minerals from the soil, convert the gas into fats, proteins, and carbohydrates. Plants, the basic ingredient of the food chain, are the foundation of the vast pyramid of life on earth.

Texas has more than five thousand species of flowering plants varying in size from herbs less than an inch tall to trees topping 150 feet. Contrast the 3-foot leaves of the sabal palm with the ½-inch leaves of Apache plume. Magnolia flowers spread 8 inches across, and giant ragweed has ⅛-inch flower clusters. Some plants occur only in swamps, whereas others prefer the torrid deserts. Why this vast diversity in nature?

Some environments provide plants with a rich source of matter and energy: water, carbon dioxide, minerals, and sunlight—the basics for the survival and growth of the organism. In the forests of East Texas, for example, trees usually have an abundance of moisture and mineral-rich soil. Other environments are limited in one or more of the basics. Each plant must compete for matter and energy with others of its species, with different species, and with those latecomers on the scene, animals. In the eastern forests, trees and shrubs compete for available sunlight. In the Chihuahuan Desert of West Texas, plants have plenty of sunlight but very little water.

Terrestrial animals, which evolved about 26 million years after the first land plants, survive by exploiting plants as a ready-made food source. The pressures of competition from other plants and from herbivores force plants to develop a wide variety of strategies for growth, reproduction and defense. By becoming specialists in using sunlight, water, and other materials available in a specific habitat, many species of plants can coexist without competing with each other.

In nature, form follows function. Virtually every feature, even the most minute characteristic, helps a plant survive. Its color, size, shape, type of flower, odor, hair structure, and chemical competition adapt it to extract energy from sunlight, to reproduce, and to ward off herbivores (plant eaters). For example, cacti have adapted to life in a harsh, water-limited environment. Their waxy epidermis reduces water loss to a fraction of that of other plants, and a network of shallow roots rapidly absorbs moisture when it is available. To further inhibit water loss, leaves have disappeared. Instead the green, fleshy stems store water and take over the job of photosynthesis. Thorns shade the plant, trap moisture, and discourage nibblers, and spectacular flowers attract insects for pollination. Superbly adapted to arid conditions, most cacti die in a matter of weeks if transplanted to waterlogged soil.

You can get a quick view of the marvelous adaptations and diversity in the plant kingdom by looking at leaves, the food factories for plants. Besides leaf shape, size, and color differences, the surface texture varies from smooth to sandpapery rough or densely fuzzy, the edge from smooth or coarsely toothed to tipped with barbs, and the blade from stiff to flexible. The plant may be evergreen or deciduous, and the leaves may have a pleasant or repugnant scent resulting from aromatic oils. All of these characteristics are factors in the survival equation. Each has its purpose.

The physical characteristics of a plant's leaf can tell you a lot about the plant's habitat. In a dense forest, large trees restrict the amount of sunlight available to the understory shrubs. Shrubs are able to compensate by producing large leaves. The larger the leaf, the more surface area is available for absorbing sunlight and carbon dioxide, essentials for photosynthesis. Larger surface area also means increased loss of moisture through the leaves (transpiration). Each day a large broadleaf tree may transpire hundreds of gallons of water. In a sunny habitat with ample moisture, large leaves can produce enough carbohydrates through photosynthesis to enable the plant to grow rapidly. In low-light conditions, such as the understory of a forest, a large leaf may be a necessity for the plant to absorb enough sunlight to survive.

On a transect going west from the moist coast of Texas, rainfall diminishes from 60 to 8 inches annually. At some point on the westward continuum, trees and shrubs with large leaves lose more water than is available for the roots to absorb. In arid conditions, particularly in areas with high summer temperatures (which increase transpiration rates), plants must find ways to decrease water loss to survive. Where sunlight is not a limiting factor, such as in West Texas, a shrub can get by with a small leaf. The smaller leaf will decrease water loss while still absorbing enough solar energy to power the food factories. But a smaller surface area also reduces the amount of carbon dioxide the leaf can absorb, which, along with the limited water budget, slows the growth rate. Generally, trees and shrubs in West Texas have smaller leaves than their wetland relatives. Blackjack oaks in East Texas have much larger leaves than those growing in the drier central part of the state. Forests in East Texas have a great diversity of oaks, some more than 100 feet tall; the arid sand counties of West Texas have thousands of square miles covered with 3-foot-tall scrub oaks.

Besides being small, the leaves of desert plants often are densely covered with hair. The fuzzy coat reduces transpiration by providing shade and by creating an insulating dead air space around the leaf. Additionally, some plants, such as ocotillo, produce leaves only during the rainy season. Usually, you would be correct in guessing that a small hairy leaf came from a medium to small tree or shrub growing in a dry habitat, and a large, shiny leaf from a tree or shrub in a moist habitat.

Plants and insects have evolved together for 200 million years, and many of their characteristics result from complex interrelationships. Numerous plants have developed poisonous compounds and physical barriers as feeding deterrents. Some plants immediately begin transporting chemicals that taste bad or contain toxic compounds to injured leaves to discourage nibbling insects. On many shrubs, the leaves, stipules, and branch tips have evolved into vicious thorns to help ward off herbivores.

One of the plant features most influenced by insects is the flower. To reproduce sexually, a plant must somehow transport its pollen between flowers and often to distant plants. Showy, sweet-smelling, colorful flowers serve one purpose: to attract insects, hummingbirds, or other animals that will carry the pollen from one flower to another. As a reward, the plant gives the animal gifts of nutritious pollen and nectar. Many flowers have evolved a combination of shape, size, and smell that attracts only one species or a small group of insects or other type of animal. Yuccas have a complex life cycle dependent on the yucca moth. Only the specific behavior of the female moth can pollinate the flower. The moth collects pollen, spreads it on the receptive stigma (part of the reproductive organs of the flower), then lays her egg in the flower's ovary. The larva hatches in the developing seedpod and feasts on some of the nutritious seeds. It tunnels its way out, burrows underground, and emerges as an adult moth the next spring as the yucca produces new flowers.

Not all flowers resort to bribing animals to ensure pollination. Some trust their fate to the fickle breezes. Ragweed releases copious amounts of pollen to waft through the air in search of a receptive flower. With no need to attract insects, these flowers are tiny and inconspicuous, and they lack petals. Airborne pollination, far from a primitive counterpart to insect pollination, is an intricately orchestrated process. In the case of pine trees and other conifers, the symmetrical shape and scales of the female cone set up a complex pattern of wind and eddy currents. The air flow channels pollen directly into the cone. The shape of the microscopic grains of pollen responds only to the particular currents created by the cones of its species. The pollen grains sail right past other trees.

Once a plant has managed to get its flowers pollinated, it faces another serious problem: how to disperse its seeds. Again, plants astound us with their resourcefulness. Seeds ride the breezes on tiny parachutes (dandelions) or sail on delicate wings (maples), float across the waters on rafts (lotuses), hitchhike on animals (cockleburs), rocket through the air from spring-loaded catapults (wild geraniums), and stow away in animals' bellies. Plants produce succulent fruits and delicious nuts for their own benefit, not ours. Animals serve the plant's purpose by eating the fruit and spreading the seeds. Some plants have conditioned humans to sow their seeds across vast acres of land. Humans even cultivate the soil and carefully eliminate competing plants and insects. Who says plants aren't ingenious?

Plants give us the necessities of life: food, textile fibers, medicines, building materials, and industrial products such as rubber, wax, oils, dyes, and soap. The diversity in the plant kingdom gives us a less concrete, but equally important, contribution. The incredible adaptations of plants to their living and nonliving surroundings can fire our imagination and increase our wonder of life and the world around us.

Humans have deciphered only a few of the amazing stories found in the plant kingdom. The next episode may start in your own backyard. This book is an introduction to the cast of characters making up the intricate plot that unfolds around us season by season, day by day. It will help you learn the names of some of the main players, the ones that are your closest neighbors. We think you will find that learning their names will in some small way make your life richer. Learning names is the first step to making friends, and making friends always enriches our lives. Friends, whether plant, animal, or human, entertain us, stimulate us, teach us, and help us understand our personal responsibility to the world around us. So take this book and start meeting some of our most interesting neighbors. ❧

4

Endangered Species— Our Disappearing Botanical Heritage

Have you ever cut down a tree, a big tree? My son and his grandfather did—a 60-foot pine for firewood. The chain saw screamed, the tree leaned, then came crashing to the ground. Pop smiled in anticipation of spending many pleasant winter evenings in front of a glowing fireplace. Like Jack the Giant Killer, my ten-year-old son stood on the massive trunk and grinned with the feeling of raw power. Something inside of us glows in satisfaction when we turn directly to nature for our daily necessities, whether it be firewood or food from a backyard garden.

Through the years, humans have methodically, and at an accelerating pace, altered the natural habitat to meet the increasing needs of a mushrooming population. Today the change is occurring so rapidly that the present landscape hardly resembles what our parents and grandparents knew. Cultivated fields and pastures have replaced the prairies that once stretched from South Texas to Canada. Satellite photos show that lumbering, agriculture, reservoirs, and population growth have destroyed more than two-thirds of the prime habitat in East Texas, which once harbored the greatest diversity of plant and wildlife in North America. Only 1 percent of the subtropical Lower Rio Grande Valley has escaped cultivation. Perhaps the greatest threat to our native plants, however, is not the cow and the plow but urban sprawl. Each year the last tree falls on tens of thousands acres as cities and suburbs spread concrete and turf grass across the countryside.

All too often city dwellers find that they must visit nature preserves to see the native plants that once provided food, fuel, clothes, and shelter for our ancestors. Our growth may cause 3,000 of the 25,000 species of trees, shrubs, herbaceous wildflowers, and other plants in the United States to disappear before your grandchildren are old enough to venture into the woods. Texas, with its assortment of ecological habitats, ranks third on the list of states with the most species of threatened and endangered plants.

The first step in preserving our threatened biological diversity is to identify the rare species of plants and animals and the habitat requirements necessary to sustain them. The Texas Parks and Wildlife Diversity Program tracks 238 rare, threatened, and endangered plant species in the state. The plant list includes 110 species that occur in fewer than five places in Texas. Of those, 55 grow in less than six locations on the planet. Some, such as Texas wild rice in the San Marcos River, grow in a single location. Another 110 species on the list grow in fewer than twenty locations globally. As Texans we have our share of rare botanical treasures.

Although some plants survive by adapting to a wide range of soil and moisture conditions (hackberry trees grow throughout the state), more exacting plants specialize in a particular set of environmental conditions. Some cacti grow only in one desert valley in West Texas. The naturally rare plants are the most susceptible to extinction. The slightest disturbance to their habitat may drive them to oblivion. The rare cacti in the Trans-Pecos face another threat: overcollecting by plant lovers and cactus rustlers from around the world. Texas has more species of cacti than all other states combined, but unlike some states that have outlawed cactus collecting, Texas has no laws specifically protecting these unusual plants. Cacti represent nearly 50 percent of the federally listed endangered plants in Texas.

The cacti and other plants in the Trans-Pecos must survive in a land of extremes. Desert basins of creosote bush and tarbush surround mile-high peaks capped with ponderosa pine and aspen. Summer temperatures often exceed 110°F, and winter temperatures may plummet from 80 to 15°F in a matter of hours. The parched arroyos may go months without a drop of rain, then overflow their banks for a few brief hours after a thunderstorm. This country may look rugged, but the ecological balance is easily upset. The Trans-Pecos has three times more species of endangered and threatened plants than any other region in Texas. The 2002 Texas Wildlife Diversity Program list includes about 30 species for Central Texas, 40 each for East and South Texas, and 138 for the Trans-Pecos.

Unwise range management has irreparably altered much of the landscape in West and Central Texas. Overgrazing by cattle, sheep, and goats has removed the vegetative cover protecting the thin layer of topsoil. Rains from torrential thunderstorms typical of the area have washed away the precious soil. Without a thick cover of native grasses and other plants, rain rapidly drains into creeks and arroyos instead of soaking in and replenishing the springs. As permanent springs and creeks dry up, many of the plants and animals dependent on them disappear.

Suppression of wildfires has altered the plant composition of Central and West Texas as much as overgrazing. Periodic burning prohibits many species of trees and shrubs from invading grassland prairie and savannahs, which are composed of fire-tolerant grasses and wildflowers. Without range fire to keep Ashe junipers (mountain cedars) in check, dense brakes of these trees have replaced the grassland savannahs of Central Texas. Pioneer species such as mesquite, prickly pear, and broomweed have claimed thousands of square miles that once supported knee- to waist-high grasses and a diversity of wildflowers.

What can you and I do to help preserve our precious botanical heritage? We may not be able to stop the advancement of shopping centers, subdivisions and industry, but we can press for city ordinances that set aside a percentage of the land as parks, greenbelts, and preserves. Think of how important expansive urban parks are to the citizens of the state's large cities: San Antonio's Brackenridge Park, Houston's Memorial Park, Fort Worth's Forest Park, Austin's Zilker Park, and the Dallas Arboretum and Botanical Garden preserve an outdoor lifestyle as much as native plants and animals. Also, we can refuse to buy wild-collected, rare plants, primarily cacti and we can encourage stronger legislation to protect our rare plants. We can support organizations that help preserve plants and their habitat. Investigate the work done by Texas Nature Conservancy, the Lady Bird Johnson Wildflower Center, Tree Folks, area nature centers and public gardens, garden clubs, Native Prairies Association of Texas, and local chapters of the Native Plant Society of Texas (http://www.npsot.org), Flora of Texas Consortium, (http://www.csdl.tamu.edu/FLORA/ftc/ftchome.htm), Sierra Club and Audubon Society. Many of these organizations now have web-sites. University Bioinformatics Working Group now has an extensive database on the Internet providing information on Texas plants and animals. (http://www.csdl.tamu.edu/FLORA/bwgprorb.htm). Two resources linked to that website provide information on threatened and endangered species—the Checklist of the Vascular Plants, and the Texas Endemics.

Education is an essential step in increasing the public's awareness of the need to preserve our botanical heritage. Many cities have nature centers and botanical gardens that present educational programs and outings for adults and children. Books and magazines can astound us with an intimate view of the mysteries abounding in nature, and field guides, such as this one, can help us develop a personal knowledge of the plants native to Texas. If we increase our awareness of the plants and animals around us, we will develop a greater value and appreciation for the delicate ecological balance necessary for sustaining life on this planet we call Earth. ❧

5

Landscaping with Native Plants

Texans have more to brag about than cattle, oil, and football teams. We can be proud of our native plants. The Lone Star State has long been famous for its spectacular array of wildflowers, but now its trees and shrubs claim a share of the glory. Until recently, only botanists and native-plant lovers recognized the benefits of landscaping with indigenous plants. Now many professional land-scapers and nurseries across the state have discovered that Texas plants survive heat, drought, and freezes much better than most imported exotics. For extensive information, refer to George Miller's book, *Landscaping with Native Plants of Texas and the Southwest*, which includes species descriptions, color photographs, and how-to chapters.

Not only do native plants survive, they thrive—and with very little maintenance. In fact, low maintenance is their strongest selling point. By selecting plants from your own area, you automatically have a landscape adapted to whatever weather extremes may occur. During ten thousand years of climatic vicissitudes, only those species adapted to the extremes in weather have survived. Once a native plant is firmly established, it needs little supplemental water, even in the driest years. Unusually frigid winters may nip tender twigs, but natives usually survive and fully recover with the next spring growth. Other than normal landscape maintenance, such as pruning, a native plant requires little attention.

Every plant has its optimum growing conditions—the temperature, humidity, moisture, exposure, and soil type with which it best thrives. The requirements of plants from your area are more likely to match the conditions of your landscape setting than those originally from some distant continent. Local plants have grown for centuries in the soil of your locale and require less fertilizer and soil conditioner to grow rapidly and remain healthy. Fewer native plants will die and have to be replaced, and the foliage and general appearance of the plants will remain attractive throughout the years.

The low-maintenance characteristic of native plants is an attractive feature for homeowners, but it is becoming essential for municipal and commercial landscaping. Large businesses allocate sizable budgets for landscaping and for maintaining plants in an attractive manner. Cities and state agencies landscape areas around buildings, in parks, and along thousands of miles of streets and highways. The State of Texas spends millions of dollars anually on landscaping. State agencies have pledged to use at least 60 percent natives in their landscaping projects. Plants that minimize water use and labor costs mean sizable savings to each of us as homeowners, business people, and taxpayers.

Most large metropolitan areas must find ways to conserve water, or their wells and reservoirs will run dry in the near future. Lawn watering accounts for nearly 50 percent of residential water consumption. Xeriscaping (*xeri*, "drought-tolerant," *scape*, "landscape") helps relieve the demand for water and reduces our personal utility expenses. In the western half of the state, drought-tolerant species should be the dominant plants of any landscape design.

Native plants offer more to homeowners and professional landscapers than low maintenance—they offer beauty and diversity. Texas plants come in all shapes, sizes, and hues. You can choose a small, medium, or large tree or shrub to fit any landscape design. Among the many species, some flower in the spring, and others in the fall. Some are evergreen, and others lose their leaves in the winter. Many shrubs have dense foliage that can be sheared and shaped, whereas others are loosely branching and best suited for informal designs. You can add color variety to your yard by choosing plants with silver or gray, woolly or lustrous, or light or dark green foliage. Our rich botanical heritage is a treasure chest for landscapers.

At least two schools of thought have developed concerning landscaping with indigenous plants. One approach simply substitutes natives for the commonly used imported species in a traditional landscape design. Native plants make attractive shade trees, foundation hedges around buildings, border hedges along walks and drives, and container plants. They can be used in any formal or informal landscape design for yards and commercial plantings, and to accent entryways, courtyards, pools, and patios.

At the other end of the spectrum is the attempt to duplicate the natural plant associations found in the wild. A yard becomes a macrocosm of nature with no hedges, shaped shrubs, or species not from the immediate area. Of course, these extremes can be modified considerably to create intermediate designs between a formal landscape and a wild one.

Planting landscape islands combines the concepts of maintaining natural plant associations and using plants to visually accent open areas. Instead of a yard of turf grass delineated with hedges and accented with a few shrubs and trees, use a mass planting, or island, of mixed species. A landscape island can be completely surrounded by the yard or curve out from a building. It can include one side of a drive or complement a corner. Group plantings provide visual interest and are flexible in design.

Regardless of where you live in the state, native plants can provide year-round beauty for your yard. Choose species that complement each other as one season fades into the next. Incorporate both spring and fall flowering species of trees and shrubs. Choose some with attractive berries in the summer and others that fruit in the winter. A combination of evergreen and deciduous plants yields dramatic results. The evergreens provide green foliage during the barren winter months. Deciduous species add shades of bright green as new spring leaves emerge, and some turn spectacular shades of scarlet in autumn. Hedges do not have to be of a single evergreen species but can combine the best nature has to offer. Group plantings can have a different accent for every season. Your yard can be a showplace throughout the year.

Fall and winter are the times to plant your landscape. Most trees and shrubs should be set out during cool months when heat and drought stresses are minimal. The plants will have all spring to grow the hardy root system necessary to stand the rigors of summer. Until the root system has had two growing seasons to develop an extensive network of feeder rootlets, the plant may need supplemental water to survive a dry summer. Once established, native plants from your area seldom require additional care. Container-grown or balled-and-burlapped plants from nurseries suffer little transplanting shock and begin vigorous growth immediately when set out. Wild-grown plants have wide-spreading root systems and require a large, intact root ball when transplanted. For this reason, a plant dug by the inexperienced usually dies.

The benefits of native-plant landscaping go beyond dollars saved and beautiful yards. Environmental education, the preservation of rare plants, and the creation of urban wildlife habitats are less tangible, but no less important, reasons for choosing natives. The plants that have sunk their roots in Texas soil since the last ice age are a part of this state's great natural heritage. Many rare or endangered plants have exceptional landscape merit and could be preserved for posterity by nursery propagation. In addition, the native plants in our yards provide food and shelter for birds and wildlife otherwise displaced by development. With planning, landscaping can help restore the biodiversity that urban sprawl destroys.

Today a growing number of people are concerned with preserving what is left of our natural environment, especially in urban areas. In most large cities the natural associations of native plants remain only in preserved sanctuaries. Why can't our neighborhoods represent the plant diversity that is found in nature or that existed before your houses were built? Our children could grow up familiar with the same plants that provided food and fiber for the Native Americans who lived in Texas for thousands of years. A stroll down our own street could be a lesson in native-plant ecology. ❧

6

Herbaceous Wildflowers

RED, PINK, BLUE, OR PURPLE FLOWERS

---

1. Wild Petunia, *Ruellia nudiflora*
Acanthaceae, Acanthus Family

Upright to sprawling perennial 1–2 feet (3–6 dm) tall; young plants hairy, stems and leaves of older plants nearly hairless; in partial shade, in various soils, in fields and along streams, roadsides; common, with several varieties. FLOWERS: Year-around; purplish-blue or white; 1½–2½ inches (4–6.5 cm) long; trumpet shaped, 5-lobed, in terminal clusters. LEAVES: Simple, opposite; blade elliptic, egg-shaped, lanceolate or oblanceolate, 2–5 inches (5–12.5 cm) long; underside prominently veined; margins toothed, undulating. RANGE: Texas; Mexico. RELATED SPECIES: 15 species in Texas, some very similar. One species, *R. drummondiana*, endemic to Central and South Texas.

2. Eryngo, *Eryngium leavenworthii*
Apiaceae, Carrot Family

Stiffly erect, prickly annual to 4 feet (1.2 m) tall; stems, leaves, and bracts usually purple; in open areas, fields, prairies; uncommon but may occur in large colonies. FLOWERS: Summer–fall; purple to red; many in cylindrical head to 1½ inches (4 cm) tall; minute flowers have blue stamens; spiny bracts above and below heads. LEAVES: deeply palmately lobed; alternate below, crowded above, with whorls below flowers; stiff, spine-tipped. RANGE: Texas to Kansas. RELATED SPECIES: Above characters distinguish this from 8 other species; most have blue or white flowers.

3. Butterfly Weed, Orange Milkweed,
Asclepias tuberosa
Asclepiadaceae, Milkweed Family

Erect perennial to 3 feet (1 m) tall; sap watery, not milky; stem and leaves hairy; in fields and open woods, in canyons, and on dunes; common in eastern part of range. FLOWERS: Spring–fall; 5 bright red or orange petals bending downward, with 5 horn-shaped appendages (hoods) protruding upward; many flowers in a loose rounded umbel. LEAVES: Simple, crowded; narrowly lanceolate or fairly broad, 1–4 inches (2.5–10 cm) long, to 1¼ inch (3 cm) wide; sessile or on stalk less than ¼ inch (6 mm) long; margins entire or wavy. FRUIT: Slender pod to 6 inches (1.5 dm) long; flat seeds attached to fluffy hairs. RANGE: Throughout Texas; eastern U.S. RELATED SPECIES: Above characters distinguish this from 35 other species. One species, *A. linearis*, is endemic to South Texas. NOTE: Milkweeds are poisonous.

4. Texas Aster, *Symphyotricum drummondii*
var. *texanum (Aster texanus)*
Asteraceae, Sunflower Family

Erect perennial 1–3 feet tall (3–9 dm) tall; upper stem branching; in partial shade, in prairies and openings in woods, along fencerows; threatened. FLOWERS: Fall, rarely to early spring; heads ½–1 inch (1.2–2.5 cm) in diameter; rays white to pale blue or lavender; disk flowers yellow or lavender. LEAVES: Simple, alternate; blades of main leaves 2–5 inches (5–12.5 cm) long, heart- to egg-shaped, leaf stalk winged; surface rough; margins toothed; upper leaves much reduced. RANGE: Texas, Arkansas, Louisiana. RELATED SPECIES: 20 species in Texas, often difficult to distinguish. See p. 119.

5. Basket Flower, Cardo del Valle,
Centaurea americana
Asteraceae, Sunflower Family

Showy annual 3–5 feet (1–1.5 m) tall; in sun, in sand or clay, in prairies and fields, along roadsides; common to abundant, may form large colonies. FLOWERS: Spring–summer; pale pink to deep lavender (rarely white), center may be creamy-yellow; solitary heads to 3½ inches (9 cm) in diameter, composed of many disk flowers; bracts enclosing head form prickly basket; stem swollen below head. LEAVES: Simple, alternate; egg-shaped to lanceolate, to 2½ inches (6 cm) long; margins entire, larger lower leaves may be toothed. RANGE: South-central U.S.; Mexico. RELATED SPECIES: Other pink members of the genus lack the prickly bracts; others have yellow flowers. Compare with thistles (*Cirsium*), below and page 91.

6. Texas Thistle, *Cirsium texanum*
Asteraceae, Sunflower Family

Prickly taprooted biennial or perennial 3–5 feet (1–1.5 m) tall; winter plant is basal rosette of large leaves; in spring sends up branching flower stalk; in dry soil, in prairies, disturbed soil and fields, along roadsides; abundant. FLOWERS: Spring-summer; pink to purple; many tubular disk flowers in rounded head to 1½ inches (4 cm) wide; spine-tipped bracts enclose head. LEAVES: Simple, alternate, also basal; basal leaves largest, 4–13 inches (1–3.2 dm) long; pinnately lobed, each lobe and tooth tipped with hard needle-like spine; leaf bottom may be woolly, white. RANGE: Oklahoma to Mexico. NOTE: Before flower buds form, the roots, young leaves, and flower stalk are sweet and edible. RELATED SPECIES: This is the most common of 10 Texas species. Compare with bull thistle, page 91, and basket flower above. One species, *C. terranigrae*, is endemic to North-Central and east Texas.

7. Mistflower, Blue Boneset,
Conoclinium coelestinum
(Eupatorium coelestinum)
Asteraceae, Sunflower Family

Bushy perennial, spreading or erect, 1–5 feet (3–15 dm) tall; stems may be purple; in sun or partial shade, moist soil, in woods and wetlands, along streams and roadsides; abundant in East Texas, uncommon further west; may form large colonies. FLOWERS: Late summer–fall; tiny blue or purple heads in small, compact clusters; disk flowers only, with conspicuous threadlike styles. LEAVES: Simple, opposite; blades triangular (sometimes egg-shaped), 1–5 inches (2.5–13 cm) long; margins toothed. RANGE: eastern and southern U.S. to Kansas. RELATED SPECIES: 2 other species in Texas.

8. Purple Coneflower, *Echinacea sanguinea*
Asteraceae, Sunflower Family

Perennial with flowers solitary on tall stalk to 3 feet (1 m) tall; in sand or gravel, in prairies and pine forests, along roadsides; common. FLOWERS: Late spring; large heads; ray flowers pink to purple, long, narrow; many disk flowers, purple or brown; heads on nearly spherical receptacle covered with stiff scales. LEAVES: Simple, alternate, most near the base; blades lanceolate, to 10 inches (2.5 dm) long; sandpaper texture; margins entire. RANGE: South-central US. RELATED SPECIES: 5 other species in Texas, some very similar.

9. Firewheel, Indian Blanket, *Gaillardia pulchella*
Asteraceae, Sunflower Family

Colorful annual to 2 feet (6 dm) tall; in sun or partial shade, in prairies and fields, and along seashore; abundant in late spring, forming large colonies. FLOWERS: Early spring–fall; red and yellow heads 1–2 inches (2.5-5 cm) wide; 6–12 ray flowers, red with yellow tips, tips 3-lobed; disk flowers yellow, red, or brownish, many, on spherical receptacle. LEAVES: Simple, alternate, also basal; oblong, elliptic, or oblanceolate, 1–3½ inches (2.5–9 cm) long; covered with stiff hairs; margins toothed, deeply lobed, or entire. RANGE: South-central and southeastern U.S.; Mexico. RELATED SPECIES: 7 species in Texas. Most have rays that are entirely red or yellow. *G. aestivalis*, endemic to Hardin Co., may be endangered.

10. Gayfeather, Blazing Star, *Liatris mucronata*
Asteraceae, Sunflower Family

Showy perennial to 3 feet (1 m) tall, from bulblike root; often with many stems from base; in dry limestone soil, on hillsides, in prairies and fields; common. FLOWERS: Late summer–fall; purple, showy; 3–6 disk flowers to ¾ inch (2 cm) long in small heads; no ray flowers; heads dense on tall spike. LEAVES: Simple, alternate; narrowly linear, 1–4 inches (2–10 cm) long (diminishing up stem), ⅟₁₆–¼ inch (1–6 mm) wide; densely covering stem; margins entire. FRUIT: Achene attached to feathery plume. RANGE: South-central U.S.; Mexico. RELATED SPECIES: 11 other species in Texas, some difficult to distinguish; 3 East Texas species are endemic and one may be endangered.

11. Skeleton Plant, *Lygodesmia texana*
Asteraceae, Sunflower Family

Slender perennial to 2 feet (6 dm) tall; stems green, smooth; in dry, rocky, usually limestone soil, prairies and hills; common. FLOWERS: Spring–fall; lavender (bluish, pink, rarely white); solitary cup-shaped head 1–2 inches (2.5–5 cm) wide, at top of wiry stalk; 8–12 ray flowers, no disk flowers. LEAVES: Simple, alternate, also basal; upper stem leafless or leaves tiny, scalelike; lower leaves and leaves of basal rosette (drying up on mature stems) ¼–8 inches (0.5–20 cm) long; linear and entire, or deeply pinnately lobed with linear segments. RANGE: Texas, Oklahoma, New Mexico, Mexico. RELATED SPECIES: Above characters distinguish this from 2 other Texas species.

12. *Palafoxia callosa*
Asteraceae, Sunflower Family

Slender annual to 2½ feet (7.5 dm) tall; predominantly on limestone soil; abundant. FLOWERS: Summer–fall; pink to nearly white heads ½ inch (1.2 cm) wide; head solitary at tip of flower stalk; several disk flowers, each with 5 deep lobes; no ray flowers. LEAVES: Simple, alternate; narrowly lanceolate to linear, 1–2½ inches (2.5–6.5 cm) long, ⅟₁₆–¼ inch (2–6 mm) wide; margins entire. RANGE: Missouri to Texas. RELATED SPECIES: 6 other species in Texas, most occurring outside the range of this species, 3 endemic to East and South Texas.

13. Western Ironweed, *Vernonia baldwinii*
Asteraceae, Sunflower Family

Hairy perennial 2–6 feet (6–20 dm) tall; stems branching in flowering portion; in fields, prairies, plains, open woods; common. FLOWERS: Summer–fall; purple to deep pink heads, ¼–½ inch (6–12 mm) long; many disk flowers per head; heads in large terminal clusters; bracts often purplish, margins and midvein dark, surface dotted with yellow resin (hand lens). LEAVES: Simple, alternate; blade elliptic or lanceolate, 3–9 inches (8–23 cm) long, ¾–2 inches (2–5 cm) wide; base narrow; margins finely toothed. RANGE: central U.S. RELATED SPECIES: *Vernonia larsenii*, in Terrell and Pecos Counties, leaves linear, top and bottom white, woolly; *V. lindheimeri*, on Edwards Plateau, leaves linear, bottom covered with white woolly hairs. Above characters should distinguish these from 5 other species in Texas, but hybrids occur.

14. Cardinal Flower, *Lobelia cardinalis*
Campanulaceae, Bluebell Family

Brilliantly colored perennial usually under 4 feet (1.2 m) tall; in shady wetlands, in or beside streams, swamps, ponds; widespread but uncommon to rare, with 4 varieties. FLOWERS: Spring–fall; scarlet (rarely white); 1–2 inches (2.5–5 cm) long; 5 petals form tube with bilateral opening; bottom lip 3-lobed, top lip with 2 narrow arms; stamen tube projecting upward; numerous flowers in slightly 1-sided raceme on upper third of plant. LEAVES: Simple, alternate; egg-shaped to lanceolate, 3–8 inches (8–20 cm) long; margins toothed. RANGE: Eastern, central, and southwestern U.S., into Mexico and eastern Canada. NOTE: Misuse of the plant for medicinal purposes caused deaths among early settlers. SIMILAR SPECIES: See Tropical Sage, page 43.

15. Dayflower, Widow's Tears, Hierba del Pollo, *Commelina erecta*
Commelinaceae, Spiderwort Family

Perennial with succulent stems to several feet long, erect to partially prostrate; in various soils, often in shade, in fields, gardens, and woods and along streams; common and weedy, with 3 varieties. FLOWERS: Spring–fall; blue; 3 petals, with the upper 2 blue, ½ to 1 inch (12–25 mm) wide, and the lower 1 smaller and translucent, whitish; flower buds enclosed by folded bract. LEAVES: Simple, alternate; grasslike, blade quite variable, narrowly linear to broadly lanceolate, to 6 inches (1.5 dm) long, with base surrounding stem; margins entire. RANGE: Southeast and central U.S. NOTE: Young leaves, flowers, and stems (but not roots) are edible. If you squeeze the bracts below the flower, a liquid oozes out, inspiring the name *widow's tears*. RELATED SPECIES: This is the most common of 5 Texas species. *Tinantia anomola*, endemic to the Edwards Plateau, looks very similar, but the bract is not folded and does not enclose the buds.

16. Spiderwort, *Tradescantia* species
Commelinaceae, Spiderwort Family

Erect or trailing perennials 4–36 inches (1–10 dm) tall; stems succulent, sometimes branching; in sun or partial shade, in prairies and woods, along stream banks; common. FLOWERS: Late winter–summer; 3 delicate purple, blue, or pink (sometimes white) petals; flowers about ½–1½ inches (1.2–4 cm) wide, in clusters; buds on drooping stems. LEAVES: Simple, alternate; linear, elliptic, or lanceolate; grasslike, sheathing the stem; margins entire. RANGE: Throughout Texas. RELATED SPECIES: 14 species in Texas, difficult to distinguish, 6 endemic. NOTE: Leaves and stems (but not roots) are edible fresh or cooked.

17. Woolly Loco, *Astragalus mollissimus*
Fabaceae, Legume Family

Perennial to 1 foot (3 dm) tall; stems prostrate, tips erect; densely hairy, hairs feltlike or silky; in sun, in dry sandy or gravelly soil; abundant, with 5 varieties. FLOWERS: Spring; purple, pea-type; in spikes on spreading stalks. LEAVES: Pinnately compound; leaflets egg-shaped to oblong, ¼–1½ inches (6–40 mm) long; margins entire. FRUIT: Fat pod, to 1 inch (2.5 cm) long. RANGE: Wyoming and Nebraska to Mexico. NOTE: Woolly loco is poisonous to livestock, especially horses. RELATED SPECIES: 26 other species in Texas.

18. Scarlet Pea, *Indigofera miniata*
Fabaceae, Legume Family

Ground-hugging vining perennial, stem prostrate or somewhat erect, to 8 inches (2 dm) tall; in sandy or calcium-rich soils, in open areas, fields and lawns, and along roadsides and beaches; inconspicuous but common, with 3 varieties. FLOWERS: Spring–fall; peachy-pink to red; pea-type about ¾ inch (2 cm) long. LEAVES: Alternate, pinnately compound with 5–9 leaflets; leaflets ¼–1 inch (7–25 mm) long, usually not opposite; margins entire. FRUIT: Straight linear pod to 1½ inches (4 cm) long. RANGE: Texas, Florida; Mexico, Cuba. RELATED SPECIES: 2 other species with stems erect and leaflets usually paired—*Indigofera lindheimeriana*, on the Edwards Plateau and *I. suffruticosa*, naturalized from the coast inland to Bastrop.

19. Texas Bluebonnet, *Lupinus texensis*
Fabaceae, Legume Family

Annual 6–24 inches (1.5–6 dm) tall; soft hairs cover plant; mostly in calcareous or clay-rich soil, spectacular displays in fields and

along roadsides; abundant. FLOWERS: Spring; fragrant; blue with white to pink spot on upper lip; pea-type, about ½ inch (1.2 cm) long; numerous flowers in dense raceme; hairy buds form silvery white tip. LEAVES: Palmately compound, alternate (basal rosettes in winter); 5 leaflets (rarely 6 or 7), oblanceolate; tips usually pointed (sometimes rounded); margins entire. FRUIT: Hairy bean pod 2–3 inches (5–8 cm) long. RANGE: Endemic to Central Texas, widely planted elsewhere. RELATED SPECIES: This is the species commonly planted along highways. *Lupinus subcarnosus* very similar, but hairs on buds not conspicuously silvery, and tips of leaflets broad, rounded. *Lupinus havardii* is the spectacular tall bluebonnet in Big Bend. All 6 Texas species of bluebonnets are designated the state flower. NOTE: Flowers yield a light green dye.

20. Sensitive Briar, *Mimosa (Schrankia)* **species**
Fabaceae, Legume Family

Low perennials with sprawling stems to 3 feet (1 m) long, usually covered with prickles; along roadsides, in lawns, fields, forests; common. FLOWERS: Spring–fall; pink to rose or purple; minute, in spherical heads to about 1 inch (2.5 cm) wide; protruding stamens create puffball effect. LEAVES: Alternate, twice pinnately compound; tiny leaflets fold up when touched; margins entire. FRUIT: Linear or oblong, prickly bean pod. RANGE: Throughout Texas, eastern and central U.S.; Mexico. RELATED SPECIES: 6 herbaceous Texas species; considered a single species by some. See woody species, p. 230.

21. Crimson Clover, *Trifolium incarnatum*
Fabaceae, Legume Family

Erect annual, 8–36 inches (2–10 dm) tall; in fields, along roadsides; abundant and weedy. FLOWERS: Spring; dark red; tiny, tubular pea-type flowers in dense terminal spikes; spikes at least twice as long as wide, on erect stalks. LEAVES: Alternate, palmately compound, with 3 leaflets; leaflets egg-shaped (widest at tip) to nearly circular, ½–1½ inches (1.2–4 cm) long; hairy; margins with tiny teeth. RANGE: Native of Europe, now widely naturalized. RELATED SPECIES: above characters distinguish this from 13 other species. See white clover, page 124.

22. Winter Vetch, *Vicia dasycarpa*
Fabaceae, Legume Family

Trailing or weakly climbing annual or short-lived perennial vine; slender stems with tendrils; fields, roadside grass, in disturbed areas; locally abundant. FLOWERS: Spring–summer; purple, rose, lavender, or white; pea-type, about ½ inch (1.5 cm) long; about 5–20 flowers, dense on a 1-sided spike. LEAVES: Alternate, pinnately compound, midrib ending in tendril; 10–20 narrow leaflets ½–1 inch (12–25 mm) long; margins entire. FRUIT: Bean pod about 1 inch (2.5 cm) long. RANGE: Native of Europe, now widely naturalized. NOTE: Winter vetch was previously a valuable cover crop for rebuilding soils. RELATED SPECIES: above characters distinguish this from most of 11 other species.

23. Mountain Pink, *Centaurium beyrichii*
Gentianaceae, Gentian Family

Showy branching annual to 1 foot (3 dm) tall, with many blossoms forming rounded mass; in sun, on dry rocky limestone hills, or in seeps on granite; locally common, rare in western part of range. FLOWERS: Late spring–summer; bright pink, usually with white center; petals form slender tube, opening out to form a 5-pointed star about ½–1 inch (1.2–2.5 cm) wide (petal lobes about as long as tube). LEAVES: Simple, opposite; linear to threadlike, ½–1½ inches (1–4 cm) long, 1⁄16–⅛ inch (1–3 mm) wide. RANGE: Texas, Arkansas. RELATED SPECIES: 5 other Texas species. Lady Bird's centaury, *C. texense*—named in honor of Texas' best-known wildflower advocate, Lady Bird Johnson—has flowers smaller than ½ inch (12 mm) wide, lobes shorter than tube. *C. glanduliferum* endemic to Brewster, Pecos, & Terrell counties.

24. Bluebells, Lira de San Pedro,
Eustoma exaltatum ssp. *russellianum*
(E. grandiflorum)
Gentianaceae, Gentian Family

Showy annual or short-lived perennial to 2 feet (6 dm) tall; along roadsides, in prairies, fields; uncommon to rare. FLOWERS: Summer–fall; bluish purple (occasionally white, pink, or yellow); large, bell-shaped, 2–4 inches (5–10 cm) wide, with 5 petals; petal lobes usually longer than 1¼ inches (3 cm). LEAVES: Simple, opposite; oblong to egg-shaped, 1–4 inches (2.5–10 cm) long; margins entire; surface smooth, with bluish waxy coating. RANGE: Nebraska to Mexico. NOTE: Available for cultivation. RELATED SPECIES: *Eustoma exaltatum* ssp. *exaltatum*—in East Texas near the coast and in South Texas, Travis County, and the Trans-Pecos—has small flowers, with petal lobes less than 1 inch (2.5 cm) long.

25. Rose Gentian, Meadow Pink,
Sabatia campestris
Gentianaceae, Gentian Family

Slender, erect annual 6–20 inches (1.5–5 dm) tall; stem branching; in sun, in moist or dry soil, in fields, prairies, openings in woods; common in some areas. FLOWERS: Spring–summer; dark rose to pink (rarely white), with star-shaped yellow center; 5 petals, each ⅜–1 inch (1–2.5 cm) long, to ⅝ inch (15 mm) wide, usually as long or longer than sepals; sepals with narrow lobes, bases forming winged tube; pistil with 2 styles, at first twisting and lying flat, later erect and uncoiling. LEAVES: Simple, opposite; blade egg-shaped, elliptic, or lanceolate, ½–1¾ inches (1.2–4.5 cm) long, clasping stem; thin, membranous; margins entire. RANGE: South-central U.S. RELATED SPECIES: Above characters distinguish this from 6 other similar species.

26. Texas Storksbill, Cranesbill,
Erodium texanum
Geraniaceae, Geranium Family

Low-growing annual or biennial; stems spreading or erect, to 1½ feet (4.5 dm) tall; in rocky or sandy soil, in open areas, fields, road-sides, prairies; common. FLOWERS: Late winter–spring; reddish purple; about 1 inch (2.5 cm) across; 5 petals. LEAVES: Simple, alternate and opposite, also basal; blade triangular or heart-shaped, to 1½ inches (4 cm) long; toothed and often palmately lobed. FRUIT: Seed attached to erect beak to 3 inches (7.5 cm) long, resembling a stork's bill; spirally twisted when mature. RANGE: Southwestern U.S. RELATED SPECIES: *Erodium botrys*, introduced in South Central Texas, has tiny flowers and deeply pinnately lobed leaves; *E. cicutarium*, in Central and West Texas, has tiny, pale lavender flowers and finely dissected, fernlike leaves. Compare with wild geranium, page 125.

27. Nama, *Nama hispidum*
Hydrophyllaceae, Waterleaf Family

Small hairy annual 4–20 inches (1–5 dm) tall; in full sun, in gravel or sandy soil, in various habitats; common. FLOWERS: Spring–summer; pink or purple, often with yellow or white center; 5 petals form small trumpet to ⅗ inch (1.5 cm) long. LEAVES: Simple, mostly alternate; linear, oblong, or widened at tip, ½–2¾ inches (1.2–7 cm) long; margins flat or rolled under, generally entire. RANGE: Southwestern U.S. RELATED SPECIES: 11 other species, some difficult to distinguish.

28. Baby Blue-eyes, *Nemophila phacelioides*
Hydrophyllaceae, Waterleaf Family

Low, many-branched annual, hugging ground or to 2 feet (6 dm) tall; in partial shade, in moist soil, in woodlands, river bottoms, prairies; common, sometimes in large colonies. FLOWERS: Spring; pale blue to purple, with white center; about 1 inch (2.5 cm) wide; 5 petals united in nearly flat disk. LEAVES: Alternate, also basal; pinnately lobed or dissected into numerous segments; hairy. RANGE: Texas, Arkansas, Oklahoma. RELATED SPECIES: *N. aphylla*, East Texas, tiny white flowers.

29. Blue Curls, *Phacelia congesta*
Hydrophyllaceae, Waterleaf Family

Erect or spreading hairy annual or biennial to 3 feet (1 m) tall; in rocky, sandy, or gravelly soil, widespread; common. FLOWERS: Spring; purple, blue, or white; 5 petals unite to form open bell ¼ inch (8 mm) wide; blossoms densely crowded on curled 1-sided raceme; 5 stamens, longer than petals. LEAVES: Alternate; deeply pinnately divided into numerous lobed and toothed segments; often sticky. RANGE: Southwestern U.S.; Mexico. RELATED SPECIES: This is the most common of 15 species, some very similar. *Phacelia pallida*, an endangered species in the Trans-Pecos, has leaf lobes not cut to the midrib. Two endemic species.

30. *Herbertia lahue* ssp. *caerulea*
   (*Alophia drummondii*)
   Iridaceae, Iris Family

Slender perennial to 1 foot (3 dm) tall, from bulb; in sand or clay, in open areas, fields, prairies; uncommon. FLOWERS:

Spring; violet, with white center often purple-spotted; about 2 inches (5 cm) wide; 6 colorful tepals, the inner 3 tiny and with blackish-purple bases. LEAVES: Alternate, also basal; linear, grasslike, about 1 foot (3 dm) long; pleated, with base sheathing stem. RANGE: Texas to Florida.

31. *Alophia drummondii (Eustylis purpurea)*
  Iridaceae, Iris Family

Slender perennial to 2½ feet (8 dm) tall, from bulb; in dry sandy soil, in prairies and open areas of pine and hardwood forests; uncommon. FLOWERS: Spring–fall; purple, with center spotted yellow and reddish-brown; about 2 inches (5 cm) in diameter; 6 colorful tepals, the outer 3 noticeably larger; opening for only a few hours. LEAVES: Alternate, also basal; linear, grasslike, about 1–2 feet (3–6 dm) long, pleated, with base sheathing stem. RANGE: Texas, possibly in surrounding states and Mexico. NOTE: Most members of iris family are toxic.

32. Celestial, *Nemastylis geminiflora*
  Iridaceae, Iris Family

Slender perennial to 1½ feet (4.5 dm) tall, from bulb; in fields, prairies, woods; widespread but often unnoticed. FLOWERS: Spring; blue (or white), with white center; 1–2¼ inches (2.5–6 cm) wide; 6 tepals, 3 yellow stamens, 6-parted style; opening for only a few hours. LEAVES: Alternate, also basal; grasslike, some taller than flower stalks; ⅛–⅜ inch (4–10 mm) wide; pleated and folded, sheathing the stem. RANGE: Texas to Missouri. RELATED SPECIES: *Nemastylis tenuis*, in Trans-Pecos mountains, has very narrow or threadlike leaves.

33. Blue-eyed Grass, *Sisyrinchium* species
Iridaceae, Iris Family

Small annual or perennial herbs, erect or sprawling, often clump-forming; most species 6–16 inches (1.5–4 dm) tall; stems flat, winged; widespread and common. FLOWERS: Most species dark blue to purple (sometimes white), many with yellow center; to ¾ inch (2 cm) wide; 6 tepals. LEAVES: Alternate, also basal; narrow, grasslike, 3–12 inches (7.5–30 cm) long; sheathing the stem. RANGE: Throughout Texas. RELATED SPECIES: 12 blue species with hybrids making identification difficult. 3 coastal species not blue—*Sisyrinchium exile*, yellow with reddish eye ring; *S. rosulatum*, white to reddish-lavender, with purple eye ring; *S. minus*, to Central Texas, flowers reddish-purple, yellow, or white.

34. Henbit, Dead Nettle, *Lamium amplexicaule*
Lamiaceae, Mint Family

Annual or biennial; often just a few inches tall or to 16 inches (4 dm); stem square; in lawns, gardens, disturbed soil; widespread and abundant. FLOWERS: Year-round, most common in late fall and winter; pink to purple; about ½ inch (1.3 cm) tall; slender tube with 2-lipped bilateral opening; bottom lip spotted; in whorls in leaf axils; buds red or purple. LEAVES: Simple, opposite; rounded blade ½–1½ inches (1–4 cm) long; surface wrinkled; upper leaves sessile; margins with rounded teeth. RANGE: Native of Europe, naturalized throughout Texas and the U.S. RELATED SPECIES: *Lamium purpureum*, in East Texas, leaves purple-tinged, all on petioles. NOTE: The young tender plants of both species are edible raw or cooked; unlike most mints, these do not have aromatic leaves.

35. Lemon Bee Balm, Horsemint,
Monarda citriodora
Lamiaceae, Mint Family

Aromatic annual or biennial to 3 feet (1 m) tall; stem square; in fields, prairies, hills; abundant, with 2 varieties. FLOWERS: Spring–summer; pink to pale purple (sometimes white), often purple-spotted; about ¾ inch (2 cm) long; two-lipped, with upper lip arching; in interrupted spikes; calyx teeth threadlike; whorls of leaflike bracts below flowers; bracts usually purple, oblong to lanceolate, with threadlike tip. LEAVES: Simple, opposite; blade lanceolate or oblong, ¾–2½ inches (2–6.5 cm) long or longer, ¼–½ inches (6–12 mm) wide; petioles to 1¼ inch (3 cm) long; margins toothed to nearly entire. RANGE: South-central U.S.; Mexico. NOTE: The volatile oil, citronellol, is used in perfumes and insect repellent. RELATED SPECIES: 6 other species, some similar. See *Monarda punctata*, page 106.

36. False Dragon's head, Obedient Plant,
Physostegia pulchella
Lamiaceae, Mint Family

Colony-forming perennial 1–5 feet (3–14 dm) tall; stem four-sided; in moist soil, along streams, in ditches, prairies, edges of swamps; common. FLOWERS: Spring–summer; reddish purple to pink, with purple streaks and spots; tubular, ¾–1¼ inch (2–3 cm) long, somewhat inflated; 2-lipped, upper lip not lobed, lower lip 3-lobed; 4 stamens; flowers in pairs on a terminal spike. LEAVES: Simple, opposite or basal; blades linear, oblong, or narrowly lanceolate or oblanceolate, 1–6 inches (2.5–15 cm) long; leathery; margins toothed to nearly entire. RANGE: Endemic to Texas. RELATED SPECIES: 6 other species in Texas, some very similar, 2 endangered. Compare with foxglove, page 131.

37. Tropical Sage, Mirto, Mejorana,
Salvia coccinea
Lamiaceae, Mint Family

Aromatic perennial 6–36 inches (1.5–10 dm) tall; stem square; stem and leaves often densely hairy; usually in sandy or clay soil, in moist thickets and woods and along streams; common in some areas. FLOWERS: Late winter–fall; scarlet (rarely pink or white); tubular, about 1 inch (2.5 cm) long; bottom lip lobed, upper lip narrow; 2 stamens. LEAVES: Simple, opposite; blade triangular to heart-shaped, 1–2½ inches (2.5–6.5 cm) long; margins toothed. RANGE: Southeastern U.S.; Mexico. NOTE: Related to culinary sage, the leaves can be used as a spice. Attractive as ornamentals. RELATED SPECIES: Red-flowered species—*Salvia penstemonoides*, an endangered endemic species on the Edwards Plateau; *S. roemeriana*, common on limestone soil in Central Texas, rare farther west; and the shrubs *S. greggii*, in South and West Texas, and *S. regla*, in the Chisos Mountains.

38. Mealy Sage, *Salvia farinacea*
Lamiaceae, Mint Family

Perennial 1–3 feet (3–10 dm) tall; stem square; in dry limestone soil; common. FLOWERS: Spring–fall; violet-blue; tubular, ½–1 inch (1.2–2.5 cm) long; lower lip larger than narrow upper lip; 2 stamens; flowers in dense whorls on tall naked stalk; calyx densely hairy, white or violet-tinged. LEAVES: Simple, opposite and clustered; blade linear, lanceolate, or egg-shaped, 1–4 inches (2.5–10 cm) long; margins toothed, may be entire on upper leaves. RANGE: Texas to New Mexico. RELATED SPECIES: Of numerous blue-flowered sages, 2 that resemble this one are widespread—*Salvia azurea*, with calyx green, and *S. lyrata*, in East Texas, with basal leaves and margins wavy or deeply pinnately lobed in spring.

39. Blue Sage, *Salvia texana*
Lamiaceae, Mint Family

Erect perennial 6–15 inches (1.5–4 dm) tall; stems square, all 4 sides covered with long and short hairs (hand lens); in dry, rocky, limestone soil; common. FLOWERS: Spring–fall; dark blue to purple, with white eye in center; about 1 inch (2.5 cm) long; tubular and 2-lipped, broad bottom lip larger than narrow upper lip; 2 stamens, calyx densely hairy. LEAVES: Simple, opposite or in whorls below flowers (usually 2–6 leaf nodes found below flower spike); blade narrow, to 2 inches (5 cm) long; margins entire or with few teeth. RANGE: Texas; Mexico. RELATED SPECIES: Of numerous blue-flowered sages, *Salvia engelmannii*, endemic to Central and N. Central Texas, is most similar, usually with 8–11 leaf nodes below the flowers and with 4 sides of stem covered with long spreading hairs, 2 sides lacking the minute hairs.

40. Skullcap, *Scutellaria drummondii*
Lamiaceae, Mint Family

Herbaceous annual 8–12 inches (2–3 dm) tall, with square stems branching from base; hairs on upper stem long, spreading (hand lens); hairs on lower stem long or minute, appressed; in moist or dry soil, on rocky hills, in prairies and thickets, and on beaches; common. FLOWERS: Late winter–summer; purple, center white and speckled with purple; ¼–½ inch (6–12 mm) long; lower lip protruding; calyx with long hairs (hand lens); top of calyx forms a tiny cap. LEAVES: Simple, opposite; blade egg-shaped, ¼–1 inch (6–25 mm) long, not aromatic; margins entire, but lower leaves may be toothed. RANGE: Texas, Oklahoma, New Mexico; Mexico. RELATED SPECIES: *Scutellaria wrightii*, a perennial from a woody base, has minute, appressed hairs on stem and calyx; 12 other species in Texas, some very similar. NOTE: Some species toxic.

41. Wild Onion, *Allium drummondii*
Liliaceae, Lily Family

Perennial to 1 foot (3 dm) tall, from bulb; smells like an onion; mainly on dry limestone soil; common. LEAVES: Basal; linear, grasslike, to ³⁄₁₆ inch (5 mm) wide. FLOWERS: Spring; white, pink, or red (rarely yellowish); 6 tepals, to ⅜ inch (1 cm) long; flowers on slender stems forming erect umbel at top of leafless stalk. FRUIT: Shiny black wrinkled seeds. RANGE: South-central U.S.; Mexico. RELATED SPECIES: An asexual form of this species has tiny bulblets forming at the tip of flower stalk. 14 other species in Texas, some very similar; flowers white, yellow, pink, purple, or red. Several endemics. Wild garlic, *Allium canadense* var. *canadense*, common in moist soils, also has bulblets. NOTE: Wild onions may be used in the same manner as domestic onions. Crow poison (*Nothoscordum bivalve*, page 126) looks very similar but is toxic and does not have the characteristic onion odor.

42. Red Trillium, Wake Robin, *Trillium gracile*
Liliaceae, Lily Family

Unusual perennial to 1 foot (3 dm) tall; in sandy soil, in pine woods; locally common, but habitat diminishing. FLOWERS: Spring; reddish brown or dark purple (sometimes yellow); 3 narrow erect petals, ¾–1¾ inches (2–4.5 cm) long; 3 green sepals the same size as petals; flower solitary, sessile. LEAVES: 3 broad leaflike bracts, sessile at base of flower; 2½–5 inches (6.5–13 cm) long; may be mottled; margins entire. RANGE: Texas, Louisiana. RELATED SPECIES: *Trillium recurvatum*, rare in Northeast Texas, has a sessile purple or yellow-green flower, with bracts on short petioles; *T. texanum*, an endangered species in northeast Texas, has a white or pink flower on a short stalk; *T. viridescens*, in Northeast Texas, has a flower and bracts that are sessile, 1¼–2½ inches (3.5–6.5) long, and greenish, yellow-green, or purplish green.

43. Wine-cup, Cowboy Rose, *Callirhoë leiocarpa*
Malvaceae, Mallow Family

Annual, 1–3 feet (4–9 dm) tall, with several stems from slender taproot; in woods, prairies, fields. FLOWERS: Spring–summer; reddish purple to pink; 5 petals form cup about 2 inches (5 cm) in diameter; stamens form column. LEAVES: Alternate; deeply palmately lobed. RANGE: Texas, Oklahoma. RELATED SPECIES: 5 other species, similar in appearance. Some are common, but *Callirhoë scabriuscula*, with sandpapery leaves, is an endangered species found in Runnels, Coke, and Mitchell counties.

44. Turk's Cap, Monocillo, *Malvaviscus arboreus*
var. *drummondii*
Malvaceae, Mallow Family

Bushy perennial usually under 4 feet (1.2 m), occasionally to 9 feet (3 m); in shade, at edges of woods, on hills, along streams; abundant. FLOWERS: Year-round, common summer–fall; bright red; ¾–1½ inches (2–4 cm) long, 5 petals twisted around each other; stamens form long tube. LEAVES: Simple, alternate; blade 2–5 inches (5–13 cm) long, may be wider than long, base heart-shaped; top surface slightly rough or with velveteen texture; margins toothed, may have 3 shallow lobes. RANGE: Florida to Texas; into Mexico and Cuba. NOTE: This and two other ornamental varieties are frequently cultivated, especially in South Texas. The flowers and fruit are edible. Flowers and leaves yield nice dyes.

45. Globe Mallow, *Sphaeralcea angustifolia*
Malvaceae, Mallow Family

Perennial 1–6 feet (3–18 dm) tall, usually not branching; covered with stellate hairs; in rocky or sandy soil, in disturbed areas; com-

mon. FLOWERS: Year-round; lavender, pink, or reddish yellow, occasionally white; ⅜–1 inch (1–2.5 cm) wide; 5 petals; stamens form a column. LEAVES: Simple, alternate, blade narrow, linear to lanceolate, 1–3½ inches (2.5–9 cm) long or longer, gray-green; margins toothed, not lobed (except for pair of shallow lobes sometimes found near base). RANGE: Southwestern U.S.; Mexico. RELATED SPECIES: 12 other species, some similar. *S. lindheimeri* is endemic to South Texas.

46. Meadow Beauty, *Rhexia mariana*
Melastomataceae, Melastoma Family

Delicate perennial to 2 feet (6 dm) tall; in moist or wet soil, in meadows, bogs, and ditches, near water; uncommon, with 2 varieties. FLOWERS: Spring–fall; pink to white (sometimes purple), 1–2 inches (2.5–5 cm) wide; 4 petals; anthers curved. LEAVES: Simple, opposite; lanceolate or elliptic, to 2½ inches (6.5 cm) long; margins finely toothed, teeth tipped with hairs. FRUIT: Tiny urn-shaped capsule with long neck. RANGE: Eastern U.S. RELATED SPECIES: 4 other species, some very similar.

47. Amelia's Sand Verbena, *Abronia ameliae*
Nyctaginaceae, Four-o'clock Family

Low hairy perennial; sticky to the touch; stems erect or spreading; in dry sandy soil, in live oak woods, and along roadsides; uncommon. FLOWERS: Spring; bright pink, fragrant, sticky; tube about 1 inch (2.5 cm) long, opening to form 5-lobed trumpet; in spherical cluster to 2 inches (5 cm) in diameter. LEAVES: Simple, opposite, 1 of the pair usually larger than the other; variable in size and shape, but blade often 2 inches (5 cm) long or longer, heart- or egg-shaped or nearly round; margins wavy. RANGE: Endemic to South Texas. RELATED SPECIES: 4 other species in Texas. *A. macrocarpa*, endemic to East Texas, is endangered.

48. Trailing Four-o'clock, Hierba de la Hormiga, *Allionia incarnata*
Nyctaginaceae, Four-o'clock Family

Trailing hairy perennial; sticky to the touch; in full sun, on dry sand, gravel, or limestone; widespread. FLOWERS: Spring–fall; pink to purple (occasionally white); though resembling single flower ½–1 inch (1.2–2.5 cm) in diameter, the 9 tepals belong to 3 separate flowers. LEAVES: Simple, opposite, with 1 of the pair larger than the other; blades egg-shaped or oblong, to 1½ inches (4 cm) long; margins wavy. RANGE: Texas to California; Mexico to Chile.

49. Devil's Bouquet, Scarlet Muskflower, *Nyctaginia capitata*
Nyctaginaceae, Four-o'clock Family

Colorful sticky perennial with low spreading branches to 20 inches (5 dm) long and perhaps only 6 inches (1.5 dm) high; whole plant hairy, sticky, and strong-smelling; in full sun, in dry sandy, loamy, or limestone soil, in fields, on hills, and along roadsides; uncommon. FLOWERS: Spring–fall; bright red to pink; 5 tepals united in funnel shaped blossom about 1 inch (2.5 cm) long; in rounded cluster, 1½–3 inches (4–8 cm) in diameter; red stamens extend beyond tepals. LEAVES: Simple, opposite; blade triangular or egg-shaped, 1–5 inches (2.5–13 cm) long; margins entire or wavy. RANGE: Southeast New Mexico to Texas and Mexico.

50. *Gaura* species
Onagraceae, Evening Primrose Family

Annuals or perennials, most under 3 feet (1 m) tall; usually in sunny habitats. FLOWERS: White to pink to red blossoms about ¾ inch (2 cm) long; 4 petals bending forward; 4 sepals bending backward; 8 stamens, conspicuous; style with cross-shaped stigma. LEAVES: Simple, alternate, also basal. RANGE: Varies with species. RELATED SPECIES: 17 species in Texas, not easily distinguished. One endemic.

51. Prairie Primrose, Amapola del Campo,
Oenothera speciosa
Onagraceae, Evening Primrose Family

Low-growing perennial often under 1 foot (3 dm) tall; in various soils, in fields, prairies, lawns, and open woods, along roadsides; abundant. FLOWERS: Early spring–summer; pink (or white), usually with yellow center; cup-shaped; 4 petals 1–1½ inches (2.5–4 cm) long, united at base to form short tube to 3/4 inch (2 cm) long; 8 stamens; stigma cross-shaped. LEAVES: Simple, alternate; blade more or less lanceolate, to 3 inches (8 cm) long; margins entire, wavy, toothed, or deeply lobed. RANGE: South-central U.S.; Mexico. RELATED SPECIES: 33 other species in Texas, most with yellow or white flowers; sometimes called buttercups, though not in the buttercup family. See fluttermill (*Oenothera macrocarpa*), page 109. NOTE: Young leaves edible. Flowers yield yellow dye.

52. Purple Wood Sorrel, *Oxalis drummondii*
Oxalidaceae, Wood Sorrel Family

Small unbranched herb to 1 foot (3 dm) tall, from bulb; in various habitats, lawns, mowed fields; common. FLOWERS: Fall; pink to violet; bell-shaped, to ¾ inch (2 cm) wide; 5 petals. LEAVES: Basal; palmately compound with 3 broadly V- or wedge-shaped leaflets, leaflets ½–2½ inches (1.2–6 cm) wide; leaves and stems hairless. FRUIT: Slender ridged capsule to ⅜ inch (1 cm) long. RANGE: Endemic to Edwards Plateau and South Texas. NOTE: The leaves and fruit can be added to salads but should not be eaten in large quantity. Oxalic acid provides the tart flavor. RELATED SPECIES: 3 other species with pink to violet flowers, some very similar. See Yellow Wood Sorrel, page 110.

53. Rose Prickly Poppy, *Argemone sanguinea*
Papaveraceae, Poppy Family

Erect, prickly annual or perennial to 4 feet (1.2 m) tall; yellow sap; in full sun, in disturbed soil, in fields, along roadsides; abundant. FLOWERS: Late winter-spring; white to rose-red or lavender; to 3½ inches (9 cm) wide; 4–6 delicate, wrinkled petals; numerous yellow to red stamens; buds prickly. LEAVES: Simple, alternate, also basal; blade to 6 inches (1.5 dm) long; veins bluish; margins pinnately lobed, armed with stiff prickles. FRUIT: Elliptic prickly capsule filled with many tiny seeds; when capsule dries, small holes at top open up, resembling a pepper shaker. RANGE: Texas; Mexico. RELATED SPECIES: 7 other species in Texas, some white or yellow. See other prickly poppies pages 110 and 129. NOTE: Sap is a skin irritant.

54. Blue Gilia, *Gilia rigidula*
Polemoniaceae, Phlox Family

Colorful perennial 4–12 inches (1–3 dm) high; in dry sandy soil or on rocky, limestone hills; uncommon, with 3 varieties. FLOWERS: Early spring–summer, again in fall; deep blue to purple, with yellow center; showy, to ⅜–1 inch (1–2.5 cm) wide; 5 petals unite into short tube and open out to flattened wheel. LEAVES: Simple, alternate, also basal; blade ⅜–1½ inches (1–4 cm) long; hairy, sometimes sticky; margins deeply divided into linear or oblong segments; segments entire or toothed. RANGE: Southwestern U.S.; Mexico. RELATED SPECIES: Above characters distinguish this from 5 other Texas species. *G. ludens* is endemic to South Texas.

55. Scarlet Gilia, Foxfire, Skyrocket,
Ipomopsis aggregata
Polemoniaceae, Phlox Family

Erect biennial to 6 feet (2 m) tall; in sun or partial shade, at high elevations, on slopes, in open woodlands; uncommon. FLOWERS: Late spring–fall; bright red or pink; slender tube ¾–1¼ inch (2–3 cm) long, opening to a 5-pointed star; yellow dots decorate interior. LEAVES: Alternate, also basal; 1–2 inches (2.5–5 cm) long, pinnately dissected into linear segments; scattered along stem, not dense; strong odor. RANGE: British Columbia to California and northern Mexico, east to Texas. RELATED SPECIES: 7 other species with white to purple flowers.

56. Standing Cypress, Texas Plume, *Ipomopsis rubra*
Polemoniaceae, Phlox Family

Spectacular biennial densely covered with feathery leaves; stems to 6 feet (2 m) tall; in sun or partial shade, in dry sandy or rocky soil, along roadsides, in fields and open woods; FLOWERS: Summer–fall; bright red; to 1½ inches (4 cm) long, funnel-shaped with 5 rounded or pointed lobes; interior mottled. LEAVES: Alternate, also basal; 1–4 inches (2.5–10 cm) long, pinnately dissected into threadlike segments. RANGE: Southeastern U.S. NOTE: A favorite of hummingbirds. At one time a common wildflower, but habitat destruction and overcollecting have eliminated many stands. RELATED SPECIES: 7 other species with white to purple flowers.

57. Drummond's Phlox, *Phlox drummondii*
Polemoniaceae, Phlox Family

Low-growing hairy annual 4–18 inches (1–4.5 dm) tall; in sandy or well-draining calcareous soil, in fields, prairies, and open woods, or roadsides; locally abundant, with 6 subspecies. FLOWERS: Late winter–spring; deep red or purple, often with dark or pale center; petals form flat 5-lobed wheel, 1 inch (2.5 cm) wide, with slender tube, ⅜–¾ inch (1–2 cm) long. LEAVES: Simple, opposite or alternate; narrow, lanceolate, 1–3 inches (2.5–8 cm) long; softly hairy or sticky; margins entire. RANGE: Eastern U.S. RELATED SPECIES: Several subspecies, one endemic to South Texas. 11 other species in Texas, several endemic; one endangered. NOTE: Named for the botanist who first documented the wildflower in the area between Gonzales and Goliad in 1834. Today more than 200 cultivated varieties are planted in gardens worldwide.

58. Water Hyacinth, *Eichhornia crassipes*
Pontederiaceae, Pickerelweed Family

Floating, aquatic; in lakes, rivers, streams, ditches, and marshes; abundant and weedy, forming large colonies. FLOWERS: Spring–summer; blue to lavender; about 2 inches (5 cm) across; 6 petals unite to form tube with bilateral funnel-shaped opening; large orange spot appears on upper petal; in spike emerging from spathe. LEAVES: Simple, alternate, also basal; blades nearly round, shiny; margins entire; spongy leafstalk and inflated leaf base buoyant. RANGE: Native of South America, now a weed worldwide. RELATED SPECIES: *Eichhornia azurea*, less common in South Texas, lacks the inflated leaf base. NOTE: Though water hyacinths have caused millions of dollars of damage by clogging waterways, the plants have many valuable uses. In Texas they are grown in sewage treatment plants for water purification.

59. Pickerelweed, *Pontederia cordata*
Pontederiaceae, Pickerelweed Family

Aquatic perennial 1–3 feet (3–10 dm) tall, rooted in mud, with leaves and flowers held above water; in marshes, swamps, wet ditches; common. FLOWERS: Summer–fall; blue or violet; small, funnel-shaped, bilateral; upper lip with 2 yellow spots; in dense spikes about 2–7 inches (5–17.5 cm) long. LEAVES: Simple, alternate, also basal; blade broad or narrow, lanceolate, triangular, or heart-shaped, to 8 inches (2 dm) long, on long petiole; margins entire. RANGE: Eastern U.S.; Prince Edward Island, Nova Scotia. NOTE: Young leaves and seeds edible.

60. Columbine, *Aquilegia canadensis*
Ranunculaceae, Buttercup Family

Perennial with delicate foliage, 1–3 feet (3–9 dm) tall; on moist, shaded limestone ledges and in canyons; uncommon. FLOWERS: Spring; red and yellow; 5 petals to 1½ inches (4 cm) long, each shaped like tiny trumpet with spur; stamens protruding. LEAVES: Compound; leaflets divided into three parts, which may be further divided into threes, with the final sections often deeply lobed. RANGE: Eastern U.S.; Canada. RELATED SPECIES. 2 species in the Trans-Pecos with yellow flowers. *A. chrysantha* var. *hinckleyana* is endemic and may be endangered.

61. Larkspur, Espuela de Caballero,
Delphinium carolinianum
(includes *D. virescens, D. vimineum***)**
Ranunculaceae, Buttercup Family

Erect perennial 1–3 feet (3–9 dm) tall; in woods, fields, and prairies, along roadsides; common, with several subspecies. FLOWERS: Spring–summer; white, blue, or purplish; ½–1¼ inch (1.2–3 cm) wide and long; 5 sepals united, form bilateral opening, with a long spur; 4 petals smaller, inside sepals. LEAVES: Alternate, deeply palmately lobed, 1–4 inches (2.5–10 cm) wide. RANGE: Eastern and central U.S. to Canada. NOTE: All parts are deadly poisonous. RELATED SPECIES: *Delphinium madrense*, from the Edwards Plateau to the eastern Trans-Pecos, has bright blue flowers; 2 ornamentals, with 2-petaled flowers, are naturalized in Central and South Texas.

62. *Agalinis edwardsiana*
Scrophulariaceae, Figwort Family

Slender branching annual 1–3 feet (3–9 dm) tall; on dry limestone soil; uncommon. FLOWERS: Late summer–fall; pink to lavender; 5-lobed, bell-shaped, to 1 inch (2.5 cm) long; throat may be white and may have red spots and yellow lines. LEAVES: Simple, opposite; linear, ¼–1½ inches (6–40 mm) long; often tinged with reddish purple (whole stem and leaf may be reddish); margins entire. RANGE: Endemic to Edwards Plateau. RELATED SPECIES: 14 other species in Texas, often difficult to distinguish. *A. navasotensis* endemic to Grimes County.

63. Indian Paintbrush, *Castilleja indivisa*
Scrophulariaceae, Figwort Family

Hairy annual 8–18 inches (2–4.5 dm) tall; in full sun, in wet sandy and caliche soil, in prairies, along roadsides; common. FLOWERS: Spring; showy bright red bracts enclose inconspicuous flowers; flowers tubular, creamy or greenish, 1 inch (2.5 cm) long; bracts and flowers form dense spike. LEAVES: simple, alternate; linear or lanceolate; 1–4 inches (2.5–10 cm) long; margins entire or with pair of narrow lobes near base. RANGE: Texas; Oklahoma. NOTE: This species commonly adds brilliant splashes of red in fields of bluebonnets planted along highways. Some species are parasitic on grasses and other plants. RELATED SPECIES: Some yellow or purple species overlap the western range of *Castilleja indivisa*. Other red species are restricted to West Texas, except *C. purpurea* var. *lindheimeri* (page 112). Several endemic species, two endangered.

1. Wild petunia,
   Ruellia nudiflora

2. Eryngo, *Eryngium
   leavenworthii*

3. Butterfly weed,
   Asclepias tuberosa

4. Texas aster,
   *Symphyotrichum
   drummondii*

5. Basket flower,
   Centaurea americana

6. Texas thistle,
   Cirsium texanum

7. Mistflower,
   Conoclinium
   coelestinum

8. Purple coneflower,
   Echinacea sanguinea

9. Firewheel,
   Gaillardia pulchella

10. Gayfeather, *Liatris mucronata*

11. Skeleton plant,
    Lygodesmia texana

12. *Palafoxia callosa*

13. Western ironweed,
    Vernonia baldwinii

14. Cardinal flower,
    Lobelia cardinalis

15. Dayflower,
    Commelina erecta

16. Spiderwort,
    Tradescantia gigantea

17. Woolly loco,
    Astragalus mollissimus
    var. *earlei*

18. Scarlet pea, *Indigofera miniata*

20. Sensitive briar, *Mimosa* species

19. Texas bluebonnet,
Lupinus texensis

21. Crimson clover,
Trifolium incarnatum

22. Winter vetch,
Vicia dasycarpa

24. Bluebells,
Eustoma exaltatum
ssp. *russellianum*

23. Mountain pink,
Centaurium beyrichii

25. Rose gentian,
    Sabatia campestris

26. Texas storksbill,
    Erodium texanum

27. Nama,
    Nama hispidum

28. Baby blue-eyes,
    *Nemophila
    phacelioides*

29. Blue curls,
    Phacelia congesta

30. *Herbertia lahue*

31. *Alophia drummondi*

32. Celestial,
    Nemastylis geminiflora

33. Blue-eyed grass,
    Sisyrinchium species

34. Henbit, *Lamium amplexicaule*

35. Lemon bee balm,
    Monarda citriodora

36. False dragon's
    head, *Physostegia pulchella*

37. Tropical sage,
    Salvia coccinea

38. Mealy sage,
    Salvia farinacea

39. Blue sage,
    Salvia texana

40. Skullcap,
    Scutellaria
    drummondii

41. Wild onion,
    Allium drummondii

42. Red trillium,
    Trillium gracile

43. Wine-cup,
    Callirhoë leiocarpa

44. Turk's cap,
    Malvaviscus arboreus
    var. *drummondii*

45. Globe mallow,
    Sphaeralcea
    angustifolia

46. Meadow beauty,
    Rhexia mariana

47. Amelia's sand
    verbena,
    Abronia ameliae

48. Trailing four-
    o'clock,
    Allionia incarnata

49. Devil's bouquet,
    Nyctaginia capitata

50. *Gaura* species

52. Purple wood sorrel,
    Oxalis drummondii

51. Prairie primrose,
    Oenothera speciosa

53. Rose prickly poppy,
    Argemone sanguinea

54. Blue gilia,
    Gilia rigidula

55. Scarlet gilia,
    Ipomopsis aggregata

56. Standing cypress,
    Ipomopsis rubra

57. Drummond's
    phlox,
    Phlox drummondii

58. Water hyacinth,
    Eichhornia crassipes

59. Pickerelweed,
    Pontederia cordata

60. Columbine,
    Aquilegia canadensis

61. Larkspur,
    Delphinium
    carolinianum

62. *Agalinis edwardsiana*

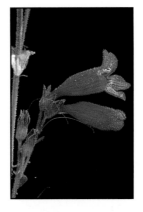

63. Indian paintbrush,
    Castilleja indivisa

64. Scarlet penstemon,
    Penstemon triflorus

65. Silverleaf
    nightshade,
    Solanum
    elaeagnifolium

66. Prairie verbena,
    Glandularia bipinnatifida

67. Texas vervain,
   Verbena halei

68. Missouri violet,
   Viola sororia var.
   missouriensis

69. Yellow star grass,
   . *Hypoxis hirsuta*

70. Huisache daisy,
   Amblyolepis setigera

71. Broomweed,
   Amphiachyris
   amoenum

72. Desert marigold,
   Baileya multiradiata

73. Damianita,
    Chrysactinia mexicana

74. Bull thistle,
    *Cirsium
    horridulum*

75. Coreopsis,
    Coreopsis tinctoria

76. Engelmann's daisy,
    *Engelmannia
    peristenia*

77. Gumweed,
    Grindelia squarrosa

78. Bitterweed, *Helenium amarum*

79. Common sunflower, *Helianthus annuus*

80. Maximilian sunflower, *Helianthus maximiliani*

81. Camphorweed, *Heterotheca subaxillaris*

82. Plains bitterweed, *Tetraneuris scaposa*

83. Texas star, *Lindheimera texana*

84. Woolly paperflower, *Psilostrophe tagetina*

85. Texas dandelion,
    Pyrrhopappus
    pauciflorus

86. Mexican hat,
    Ratibida columnaris

87. Brown-eyed Susan,
    Rudbeckia hirta

88. Texas groundsel,
    Senecio ampullaceus

89. Goldenrod,
    Solidago species

90. Sow Thistle,
    Sonchus oleraceus

91. Dandelion,
    Taraxacum officinale

92. Greenthread,
    Thelesperma filifolium

93. Cowpen daisy,
    Verbesina encelioides

94. Goldeneye, *Viguiera dentata*

95. Skeleton-leaf goldeneye,
    Viguiera stenoloba

96. *Wedelia texana*

97. Sleepy daisy, *Xanthisma texanum*

98. Zinnia, *Zinnia grandiflora*

99. Fringed puccoon, *Lithospermum incisum*

100. Bladderpod, *Lesquerella argyraea*

101. Plains wild indigo, *Baptisia bracteata*

102. Partridge pea, *Chamaecrista fasciculata*

103. *Senna
    lindheimeriana*

104. Two-leaved senna,
    Senna roemeriana

105. Scrambled eggs,
    Corydalis species

106. Spotted bee balm,
    Monarda punctata

107. Golden flax,
    Linum berlandieri
    var. *berlandieri*

108. Rock-nettle, *Eucnide bartonioides*

109. Velvetleaf mallow,
  Allowissadula
  holosericea

110. Yellow lotus,
  Nelumbo lutea

111. Spatterdock,
  Nuphar lutea
  subsp.
  advena

112. Sundrops,
  Calylophus
  berlandieri

113. Water primrose,
  Ludwigia
  uruguayensis

114. Fluttermill,
  Oenothera
  macrocarpa

115. Yellow wood sorrel,
     Oxalis stricta

116. Yellow prickly poppy,
     Argemone mexicana

117. Large buttercup,
     Ranunculus macranthus

118. Pitcher plant,
     Sarracenia alata

119. Indian paintbrush,
     Castilleja purpurea
     var. *lindheimeri*

120. Mullein,
     Verbascum thapsus

121. Tomatillo,
    Physalis cinerascens

122. Buffalo bur,
    Solanum rostratum

123. Water-willow,
    Justicia americana

124. Arrowhead,
    Sagittaria longiloba

125. Rain lily, *Cooperia
    drummondii*

126. Spider lily,
    *Hymenocallis
    liriosme*

127. Water hemlock,
     Cicuta maculata

128. Poison hemlock,
     Conium maculatum

129. Queen Anne's lace,
     Daucus carota

130. Beggar's ticks,
     Torilis arvensis

131. Green milkweed,
     Asclepias asperula

132. Yarrow, *Achillea
     millefolium*

133. Lazy daisy, *Aphanostephus skirrhobasis*

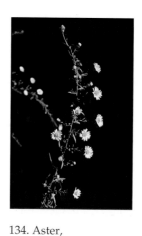

134. Aster,
Symphyotrichum
species

135. Fleabane,
Erigeron species

136. Blackfoot daisy,
Melampodium
leucanthum

137. Compass plant,
Silphium
albiflorum

138. Frostweed,
Verbesina virginica

139. Mayapple,
*Podophyllum
peltatum*

140. Shepherd's purse,
*Capsella
bursa-pastoris*

141. Peppergrass,
*Lepidium
virginicum*

143. Chickweed, *Stellaria media*

142. Clammyweed,
*Polanisia
dodecandra*

144. Snow-on-the-prairie,
Euphorbia bicolor

145. Bull nettle,
     Cnidoscolus texanus

146. White clover,
     Trifolium repens

147. Wild geranium,
     Geranium carolinianum

148. Horehound,
     *Marrubium
     vulgare*

149. Crow poison,
     *Nothoscordum
     bivalve*

150. Death camas,
     *Zigadenus
     nuttallii*

151. White stickleaf,
Mentzelia nuda

152. Devil's claw,
*Proboscidea
louisianica*

153. White water-lily,
Nymphaea odorata

154. Ladies' tresses,
Spiranthes cernua

155. White prickly
poppy,
Argemone albiflora
subsp. *texana*

156. Small pokeweed,
Rivina humilis

157. Heller's plantain,
Plantago helleri

158. Anemone,
*Anemone
berlandieri*

159. Bedstraw,
Galium aparine

160. Lizard's tail,
Saururus cernuus

161. Foxglove,
Penstemon cobaea

162. Jimsonweed, *Datura wrightii*

163. Poison Ivy,
      *Toxicodendron
      radicans*

164. Milkweed vine,
      Matelea reticulata

165. Cross vine, *Bignonia capreolata*

166. Trumpet creeper,
      Campsis radicans

167. Japanese honeysuckle,
      Lonicera japonica

168. Coral
      honeysuckle,
      *Lonicera
      sempervirens*

169. Dodder,
     Cuscuta species

170. Wild morning
     glory, *Ipomoea
     trichocarpa*

171. Wood rose,
     Merremia dissecta

172. Buffalo gourd,
     *Cucurbita
     foetidissima*

173. Balsam gourd,
     *Ibervillea
     lindheimeri*

174. Chinese wisteria,
     Wisteria sinensis

175. Krameria,
     Krameria lanceolata

176. Greenbriar,
     Smilax bona-nox

177. Carolina
     jessamine,
     Gelsemium
     sempervirens

178. Snailseed,
     Cocculus carolinus

179. Yellow passionflower, *Passiflora affinis*

180. Passionflower, *Passiflora incarnata*

181. Old man's beard,
     Clematis drummondii

182. Purple leatherflower,
     Clematis pitcheri

183. Scarlet leatherflower,
     Clematis texensis

184. Rattan vine,
     *Berchemia
     scandens*

185. Southern dewberry,
     Rubus riograndis

186. Balloon vine,
     *Cardiospermum
     halicacabum*

188. Peppervine,
    Ampelopsis arborea

187. Snapdragon vine,
    *Maurandella
    antirrhiniflora*

190. Cow-itch vine,
    Cissus trifoliata

191. Virginia creeper,
    *Parthenocissus
    quinquefolia*

189. Heartleaf
    ampelopsis,
    Ampelopsis cordata

192. Mustang grape,
    Vitis mustangensis

64. Scarlet Penstemon, *Penstemon triflorus*
Scrophulariaceae, Figwort Family

Slender upright perennial to 2 feet (6 dm) tall; stem coated with minute hairs (hand lens); in dry limestone soil on rocky slopes; uncommon. FLOWERS: spring; scarlet to pink; funnel-shaped, about 1¼ inch (3 cm) long, with 5-lobed opening; hairy; dark lines on inside; sterile stamen not conspicuously hairy. LEAVES: Simple, opposite; blade oblong, lanceolate, or oblanceolate, 1½–3 inches (4–7.5 cm) long at midstem; margins toothed to entire. RANGE: Endemic to Edwards Plateau. RELATED SPECIES: Above characters distinguish this from 22 other species, some rare, several endemic. See foxglove, page 131.

65. Silverleaf Nightshade, Trompillo,
Solanum elaeagnifolium
Solanaceae, Nightshade Family

Branching perennial 1–3 feet (3–10 dm) tall; small needlelike prickles on stem, sometimes on leaves; in sun, in dry disturbed soils; abundant. FLOWERS: Spring–fall; blue to purple, rarely white; about 1 inch (2.5 cm) wide; 5-lobed; yellow stamens protruding. LEAVES: Simple, alternate; blade narrow, 1½–6 inches (4–15 cm) long; leaf bottom and stems silvery or gray-green, covered with white hairs; margins entire or wavy. FRUIT: Yellow to black berry ½–¾ inch (1.2–2 cm) in diameter. RANGE: South-central U.S.; Mexico. NOTE: berries are poisonous but were used by Native Americans for rennet in making cheese. RELATED SPECIES: *Solanum dimidiatum* has broad, dark green leaves with shallow or deep lobes; 24 species in Texas, most white- or yellow-flowered. See *S. rostratum*, p. 113.

66. Prairie Verbena, *Glandularia bipinnatifida* (*Verbena bipinnatifida*) Verbenaceae, Verbena Family

Low hairy perennial usually under 1 foot (3 dm) tall; stems square, partly reclining, in sun, in sandy or calcareous soil, in prairies and fields, along streams and roadsides; abundant, often forming dense colonies. FLOWERS: nearly year-round, mainly in spring; pink to purple; ¼–½ inch (6–12 mm) wide; short tube with 5 lobes, tips of lobes indented, tube slightly longer than calyx; flowers dense in compressed or elongated spike at top of upright stalk. LEAVES: Opposite; 1–2½ inches (2.5–6 cm) long, pinnately divided into many narrow segments. RANGE: Alabama to Arizona, South Dakota to Mexico. RELATED SPECIES: 12 other species, some very similar. NOTE: Flowers and leaves yield nice dyes.

67. Texas Vervain, *Verbena halei* Verbenaceae, Verbena Family

Slender, erect, branching perennial to 2½ feet (7 dm) tall; stems square; in sandy or calcareous soil, in fields, prairies, and woods, along roadsides; abundant and widespread. FLOWERS: Year-round; blue or lavender (pink or white forms found in South Texas); ⅛–¼ inch (3–7 mm) wide; short tube with 5 lobes, tips of lobes notched; flowers numerous on long, slender spike. LEAVES: opposite; lower leaves broad, to 4 inches (1 dm) long, toothed to deeply lobed, on long stems; leaves diminishing upward, becoming linear; coated with rough hairs. RANGE: Texas to Alabama; Mexico. RELATED SPECIES: 18 other species, some difficult to distinguish.

68. Missouri Violet, *Viola sororia* **var.** *missouriensis*
Violaceae, Violet Family

Low perennial to 10 inches (2.5 dm) tall; in moist woods along rivers and streams; common. FLOWERS: Late winter–spring; light blue; 5 petals, bilateral; flower solitary on erect, leafless stalk. LEAVES: Simple, basal; blade heart-shaped, 1–3 inches (2.5–7.5 cm) long and wide, on stems to 8 inches (2 dm) long; margins toothed or entire. RANGE: Central U.S. RELATED SPECIES: 18 species in Texas, most restricted to East Texas, some with yellow flowers. *V. guadalupensis*, endemic, may be endangered. NOTE: Young leaves and flowers edible and high in vitamins.

YELLOW FLOWERS

---

69. Yellow Star Grass, *Hypoxis hirsuta*
Amaryllidaceae, Amaryllis Family

Perennial usually under 1 foot (3 dm) tall; in prairies, fields, open woods; common. FLOWERS: Spring; yellow; ½–¾ inch (1.2–2 cm) wide; star-shaped with 6 hairy tepals. LEAVES: Simple, basal; linear, grasslike, 4–24 inches (1–6 dm) long. RANGE: Eastern U.S.; Canada. RELATED SPECIES: 3 other species, all in East Texas, similar but rare or uncommon.

70. Huisache Daisy, *Amblyolepis setigera*
Asteraceae, Sunflower Family

Annual to 1½ feet (4.5 dm) tall; in full sun, in limestone or sandy loam soil, along roadsides, in prairies, on slopes; abundant.

FLOWERS: Winter–spring; aromatic; yellow; heads about 1½ inches (4 cm) in diameter, solitary at top of tall stalk; 8–10 large yellow ray flowers, each with 3 or 4 notches; many yellow disk flowers. LEAVES: Simple, alternate on stem, clustered at base; egg-shaped, oblong, oblanceolate, or lanceolate, 2–4 inches (5–10 cm) long; leaf base clasping the stem; margins entire, often clothed with long hairs. RANGE: Texas; Mexico.

71. Broomweed, *Amphiachyris amoenum*
(Xanthocephalum dracunculoides)
Asteraceae, Sunflower Family

Annual weed 6–36 inches (1.5–9 dm) tall; slender branches form rounded loose mass in upper part; in dry caliche soil, in overgrazed fields; abundant. FLOWERS: Summer–fall; yellow; heads ¼–⅝ inch (6–16 mm) wide, scattered over plant, not in dense clusters; 7–15 ray flowers, ⅛–¼ inch (4–7 mm) long; numerous disk flowers. LEAVES: Simple, alternate; linear or narrowly lanceolate, ¼–2 inches (6–50 mm) long; margins entire. RANGE: nearly throughout Texas; Oklahoma. NOTES: This plant is toxic to livestock. Flowers make a nice dye. RELATED SPECIES: *Amphiachyris dracunculoides* has heads in dense clusters and smaller rays. *Gutierrezia* and *Xanthocephalum* species are also similar.

72. Desert Marigold, *Baileya multiradiata*
Asteraceae, Sunflower Family

Annual or weak perennial to 1 foot (3 dm) tall; in sandy or rocky desert soil, along roadsides; abundant in Trans-Pecos. FLOWERS: Year-round, most common in spring; bright yellow heads to 2 inches (5 cm) in diameter; heads solitary at top of woolly white stem; 25–50 ray flowers about ⅜ inch (1 cm) long, with 3 teeth at tip; many disk flowers. LEAVES: Simple, alternate;

½–3 inches (1.2–7.5 cm) long; covered with woolly white hairs; margins lobed or toothed, or entire near top. Range; California to Texas; Mexico.

73. Damianita, *Chrysactinia mexicana*
Asteraceae, Sunflower Family

Fragrant, shrubby, much-branched perennial to 16 inches (4 dm) tall; in dry, sunny, rocky limestone soil; uncommon, often forming rounded mass with many flowers. FLOWERS: Spring–fall; bright yellow heads, to 1 inch (2.5 cm) wide, solitary on stalk often held above leaves; usually 12 ray flowers and numerous disk flowers. LEAVES: Simple, alternate or opposite, crowded; linear, usually less than ½ inch (1.2 cm) long, fragrant. RANGE: Texas, New Mexico; Mexico.

74. Bull thistle, *Cirsium horridulum*
Asteraceae, Sunflower Family

Spiny annual or biennial 1–6 feet (3–20 dm) tall; winter plant consists of basal rosette of large leaves; in spring, sends up thick flower stalk; in fields and open sandy areas, along roadsides; scattered, locally abundant. FLOWERS: Spring; yellow or pink to purple; many tubular disk flowers in large rounded head 2–3 inches (5–8 cm) wide; spine-tipped bracts enclose head. LEAVES: Simple, alternate on stem; basal leaves largest, to 2 feet (6 dm) long; pinnately lobed, each lobe and tooth tipped with long, needlelike spines. RANGE: Eastern U.S. near coast. NOTE: before flower buds form, the taproot, young leaves, and succulent flower stalk are sweet and edible. RELATED SPECIES: This is the only *Cirsium* with yellow flowers. See Texas thistle, (page 29).

75. Coreopsis, Golden Wave, Tickseed,
Coreopsis tinctoria
Asteraceae, Sunflower Family

Slender annual 1–4 feet (3–12 dm) tall; in sun or partial shade, usually in moist sandy soil but tolerates variety of soils, in fields and prairies, along roadsides; common to abundant. FLOWERS: Spring, occasionally to fall; heads to 1¼ inches (3 cm) wide; usually 7 or 8 ray flowers, bright yellow with reddish brown base (sometimes entirely yellow or red), and 3- or 4- toothed tip; dark reddish brown disk flowers. LEAVES: Opposite; mostly much divided into linear or lanceolate (or elliptic to obovate) segments. RANGE: Western half of U.S. and Canada. NOTE: The flowers yield yellow dye for wool. RELATED SPECIES: 12 species, some difficult to distinguish. Compare with greenthread, page 100. Two species endemic.

76. Engelmann's Daisy, Cut-leaf Daisy,
Engelmannia peristenia (E. pinnatifida)
Asteraceae, Sunflower Family

Clump-forming perennial to 3 feet (1 m) tall; in open areas, fields, and prairies, along roadsides; widespread, abundant in Central Texas. FLOWERS: Spring–summer; yellow heads 1–2 inches (2.5–5 cm) in diameter; often with 8 ray flowers that curl under when old; numerous disk flowers. LEAVES: Simple, alternate, also basal; lower leaves 6–12 inches (1.5–3 dm) long; leaves and stems covered with stiff hairs; margins deeply pinnately lobed. RANGE: Central and southwest U.S.; Mexico.

77. **Gumweed,** *Grindelia squarrosa*
Asteraceae, Sunflower Family

Gummy taprooted annual to 3 feet (1 m) tall; in sun, in dry sandy or limestone soil; common, with 2 varieties, with or without ray flowers. FLOWERS: Summer–fall; yellow heads about ¾ inch (2 cm) in diameter, sticky to the touch; ray flowers usually absent; many disk flowers; below flower head, tips of green bracts curve outward. LEAVES: Simple, alternate; egg-shaped to oblong, about 1 inch (2.5 cm) long at midstem, diminishing in size up the stem; stiff, sticky and aromatic, margins sharply toothed. RANGE: Most of U.S. RELATED SPECIES: 7 other species in Texas, 2 endangered, one endemic. Most have obvious ray flowers. NOTE: Sap useful for treating poison ivy rash in early stages.

78. **Bitterweed,** *Helenium amarum*
Asteraceae, Sunflower Family

Small clump-forming annual 4–24 inches (1–6 dm) tall; in sandy, limestone, or loamy soil, in open areas; widespread and common. FLOWERS: Spring–fall; yellow heads about 1 inch (2.5 cm) in diameter; ray flowers yellow, usually 8, with 3-lobed tips; disk flowers yellow (var. *amarum*) or reddish brown (var. *badium*); spherical receptacle. LEAVES: Simple, alternate, crowded; linear, threadlike; lower leaves may be wider and lobed; strong-smelling. RANGE: Southeastern U.S. NOTE: Bitterweed is toxic to livestock. RELATED SPECIES: 7 other species, distinguished by having leaf bases that form wings on their stems. One endemic species.

79. Common Sunflower, Mirasol,
Helianthus annuus
Asteraceae, Sunflower Family

Large annual to 8 feet (2.5 m) tall; rough hairs on leaves and stems; in fields and open dry disturbed areas; along roadsides and railroad tracks; abundant in much of range, often covering large fields; 3 subspecies. FLOWERS: Spring–fall; yellow heads 2–5 inches (5–12 cm) in diameter; 20–25 yellow ray flowers; disk flowers reddish or brown. LEAVES: Simple, usually alternate; blade egg-shaped or triangular, 2–12 inches (5–30 cm) long; on long stem; margins toothed. RANGE: Throughout Texas and U.S. NOTE: The cultivated sunflower was derived from this species. RELATED SPECIES: This is by far the most commonly encountered of 18 Texas species. Of 4 endemic species, one may be endangered.

80. Maximilian Sunflower,
Helianthus maximiliani
Asteraceae, Sunflower Family

Large perennial to 10 feet (3 m) tall; rough hairs on leaves and stems; in prairies, ditches, and fields, along roadsides; abundant in some areas, forming colonies. FLOWERS: Late summer–fall; yellow heads 2–4 inches (5–10 cm) in diameter; both ray and disk flowers yellow. LEAVES: Simple, alternate; blade narrow, lanceolate, 4–10 inches (1–2.5 dm) long, sessile; margins entire or sometimes finely toothed. RANGE: Central to southeastern U.S.; southern Canada. NOTE: This relative of the cultivated sunflower also has edible seeds and deserves cultivation. RELATED SPECIES: 18 species in Texas.

81. Camphorweed, Golden Aster,
Heterotheca subaxillaris (H. latifolia)
Asteraceae, Sunflower Family

Aromatic weedy annual 1–6 feet (3–20 dm) tall; on sandy or alluvial soil, in open dry areas; widespread and abundant. FLOWERS: Late spring–fall; yellow heads to 1 inch (2.5 cm) in diameter; both ray and disk flowers yellow. LEAVES: Simple, alternate; blade egg-shaped or elliptic, most under 1½ inches (4 cm), longer near base; strong aroma of camphor; rough, sticky hairs cover leaves; upper leaves sessile, margins toothed or entire. RANGE: Throughout Texas and southwestern U.S.; also expanding into southeastern U.S.; Mexico. RELATED SPECIES: 6 species in Texas; leaf aroma and rough texture distinguish this species from most others.

82. Plains Bitterweed, *Tetraneuris scaposa*
(Hymenoxys scaposa)
Asteraceae, Sunflower Family

Slender perennial often less than 1 foot (3 dm) tall; in calcareous or sandy soil, in dry sunny areas, prairies, plains, mountains; common, with 4 varieties. FLOWERS: Nearly year-round; yellow heads 1–2 inches (2.5–5 cm) in diameter; numerous ray and disk flowers; single head at top of long stalk. LEAVES: Simple, alternate, only on lower part of stem or all basal; narrowly lanceolate, oblanceolate, or linear, 1–5 inches (2.5–13) cm long, ⅟₁₆–⅝ inch (1–14 mm) wide; usually hairy; margins entire or with a few short lobes. RANGE: South–central and southwestern U.S.; Mexico. RELATED SPECIES: 3 other species, some similar. NOTE: Flowers make an excellent dye.

83. **Texas Star,** *Lindheimera texana*
Asteraceae, Sunflower Family

Hairy annual usually less than 1 foot (3 dm) tall; in clay or calcareous soil, in prairies, fields, open woods; common. FLOWERS: Early spring, sometimes again in fall; yellow heads 1–1½ inches (2.5–4 cm) wide, at top of stem; 4 or 5 large ray flowers; several tiny disk flowers. LEAVES: Simple, alternate, opposite or basal; blade narrow to broad, elliptic, lanceolate, oblanceolate, or egg-shaped, 1–5 inches (2.5–12.5 cm) long; margins toothed or entire. RANGE: Texas; Mexico.

84. **Woolly Paperflower,** *Psilostrophe tagetina*
   (includes P. villosa)
Asteraceae, Sunflower Family

Branching hairy perennial 4–24 inches (1–6 dm) tall; usually covered with white, woolly hairs; in rocky soils, hills, plains, or sand dunes; common. FLOWERS: Year-round; small yellow heads; 3–5 ray flowers, ⅛–½ inch (3–12 mm) long, with 3 lobes; several disk flowers; heads in dense clusters. LEAVES: Simple, alternate, also basal; narrowly oblanceolate to egg-shaped, ½–4 inches (1.2–10 cm) long; margins entire, sometimes lobed. RANGE: Southwestern U.S.; Mexico. RELATED SPECIES: *Psilostrophe gnaphalodes* has achenes covered with long, soft hairs and heads sessile or on stalks less than ³⁄₁₆ (5 mm) long. *Zinnia grandiflora* (page 102) has roundish rays.

85. **Texas Dandelion,** *Pyrrhopappus pauciflorus*
   (P. multicaulis)
Asteraceae, Sunflower Family

Erect taprooted annual 8–30 inches (2–7.5 dm) tall; stem branching; in lawns, along roadsides, fields, and prairies; common, with 2

varieties. FLOWERS: Late winter-summer; yellow heads to 2 inches (5 cm) wide; numerous ray flowers; dark stamen tubes; disk flowers absent. LEAVES: Simple, alternate or mostly basal; 1–8 inches (2.5–20 cm) long; margins toothed or deeply pinnately lobed. RANGE: Texas, Oklahoma; Mexico. RELATED SPECIES: Not always distinguishable from the 2 other species in Texas. Dandelions (page 99) similar but have all basal leaves, no branching stem, and no dark stamen tubes.

86. Mexican Hat, *Ratibida columnifera*
   (R. columnaris)
   Asteraceae, Sunflower Family

Erect branching biennial or perennial 1–4 feet (3–10 dm) tall; usually in dry limestone soil, in fields, and prairies, along roadsides; abundant. FLOWERS: Spring–fall; heads yellow and reddish brown, solitary on long stalk, held above leaves; ray flowers drooping and entirely yellow, yellow with reddish brown base, or entirely reddish brown; brown disk flowers on green cylindrical receptacle ½–2 inches (1.2–5 cm) tall. LEAVES: Alternate, dissected into many linear to narrowly lanceolate segments; stems leafy. RANGE: Central U.S.; Mexico. RELATED SPECIES: *Ratibida peduncularis*, in East Texas and along the coast, has leaves crowded near base, and segments not usually linear; *R. tagetes*, in West Texas, has small spherical or oblong heads, and rays less than ¼ inch (6 mm) long.

87. Brown-eyed Susan, *Rudbeckia hirta*
   Asteraceae, Sunflower Family

Erect annul or short lived perennial 1–3 feet (3–9 dm) tall; stems and leaves rough, hairy; in open woods, fields, prairies; roadsides; common, with 2 varieties. FLOWERS: Spring–fall; heads 2–3 inches (5–7.5 cm) across, solitary on long stalk; yellow ray flowers, usually with reddish brown spot near base; dark brown disk flowers dense on purplish brown cone, ⅜–1½ inches (1–4 cm) tall. LEAVES: Simple, alternate; shape variable but usually elliptic or lanceolate, to 7 inches (18 cm) long but generally under 2 inches (5 cm); sessile or on short stalk; margins entire or somewhat toothed. RANGE: Canada to Mexico. RELATED SPECIES: 12 other species in Texas, some similar.

88. Texas Groundsel, *Senecio ampullaceus*
Asteraceae, Sunflower Family

Erect, single-stemmed annual 1–2½ feet (3–8 dm) tall; stems and leaves usually covered with white weblike hairs or hairless; in sunny, sandy soils; often forming large colonies. FLOWERS: Spring; yellow heads about 1 inch (2.5 cm) wide; usually 8 ray flowers, numerous disk flowers; numerous heads in terminal cluster. LEAVES: Simple, alternate; lanceolate, lower leaves to 6 inches (1.5 dm) long, upper leaves small; clasping the stem; margins toothed or entire, never lobed. RANGE: Endemic to eastern half of Texas. RELATED SPECIES: 10 other species, most with lobed leaves.

89. Goldenrod, *Solidago* **species**
Asteraceae, Sunflower Family

Erect perennials; stems topped with showy golden flower heads; some species abundant along roadsides. FLOWERS: Late summer–fall; tiny yellow heads in dense clusters. LEAVES: Simple, alternate; usually lanceolate, oblanceolate, or elliptic; sessile; leaves often dense on stem; margins entire or toothed. RANGE: Varies with species. RELATED SPECIES: 24 species in Texas, difficult to distinguish. NOTE: Flowers make excellent dyes.

90. Sow Thistle, *Sonchus oleraceus*
Asteraceae, Sunflower Family

Weedy annual 6–36 inches (1.5–10 dm) or taller; sap milky; stem leafy; in disturbed soil, lawns, gardens; widespread and abundant. FLOWERS: Nearly year-round; yellow heads about 1 inch (2.5 cm) wide; numerous ray flowers, no disk flowers. LEAVES: Simple, alternate; lower leaves deeply pinnately lobed; upper leaves lobed or not, broadly lanceolate, clasping the stem; margins with softly spiny teeth. FRUIT: Achene attached to fluffy hairs; achenes to 4 times longer than wide. RANGE: Introduced from Europe, now naturalized throughout Texas and much of U.S. RELATED SPECIES: *Sonchus asper* is nearly indistinguishable; with much spinier leaves and with achene to 2.5 times longer than wide. Wild lettuce (*Lactuca* species) is very similar, with yellow, white, blue, or lavender flowers. Dandelions (see below) have all basal leaves.

91. Dandelion, *Taraxacum officinale*
Asteraceae, Sunflower Family

Weedy taprooted annual or perennial with all leaves in basal rosette; persistent invader in disturbed ground, in lawns and gardens, and along roadsides; abundant. FLOWERS: nearly year-round; yellow head about 1–1½ inches (2.5–4 cm) wide; many ray flowers, no disk flowers; head solitary at top of leafless stalk. LEAVES: All basal; 2–8 inches (5–22.5 cm) long or longer; pinnately lobed. FRUIT: Ball of achenes attached to fluffy hairs that resemble parachutes. RANGE: native of Europe, now widely naturalized throughout Texas and U.S. NOTE: The young leaves are highly nutritious; cook in several changes of water to remove bitterness. SIMILAR SPECIES: Compare with Texas dandelion, page 96, and sow thistle, above.

92. Greenthread, *Thelesperma filifolium*
Asteraceae, Sunflower Family

Branching annual or perennial 6–30 inches (1.5–7.5 dm) tall; stems leafy; in sunny dry soil, on plains, prairies, hills, beaches; abundant in much of range, with 2 varieties. FLOWERS: Nearly year-round, mainly late spring; yellow heads on tall stalks; 8 yellow ray flowers to 1 inch (2.5 cm) long, tips 3-toothed; numerous brown or yellow disk flowers; buds nodding; inner row of bracts united at base and lined with thin, papery membranes; outer row of linear bracts, ⅛–½ inch (3–12 mm) long. LEAVES: Opposite, dissected into linear, threadlike, or oblanceolate segments. RANGE: South-central U.S.; Mexico. NOTE: The dried leaves and flowers can be used for tea. RELATED SPECIES: 8 other species in Texas, some very similar, 4 endemic. Compare with *Coreopsis*, (see page 92).

93. Cowpen Daisy, Golden Crownbeard,
Verbesina encelioides
Asteraceae, Sunflower Family

Branching or bushy annual to 5 feet (1.5 m) tall; unpleasant aroma; in disturbed soils, in fields, barnyards, abundant. FLOWERS: Nearly year-round; yellow heads 1–2½ inches (2.5–6.5 cm) wide; usually 12 ray flowers, each tipped with 3 deep teeth; numerous disk flowers. LEAVES: Simple, opposite or alternate; blade of midstem leaves broad, triangular, egg-shaped, or lanceolate, with winged petiole; lower leaves to 4 inches (1 dm) long, size diminishing upward; surfaces hairy, underside gray-green or whitish; margins with coarse teeth. RANGE: Texas to Florida. RELATED SPECIES: Above characters distinguish this from 7 other species. *V. lindheimeri* is endemic to eastern Edwards Plateau.

94. Goldeneye, *Viguiera dentata*
Asteraceae, Sunflower Family

Bushy perennial, 3–6 feet (1–2 m) tall, with numerous flowers; stems woody; in dry limestone soil, in fields, hills, openings in woods; locally abundant. FLOWERS: Fall; golden yellow heads ¾–1½ inch (2–4 cm) wide; 10–14 ray flowers with notched tip, numerous disk flowers. LEAVES: Simple, opposite and alternate; blade egg-shaped to broadly lanceolate, to 6 inches (1.5 dm), but usually 1–3 inches (2.5–8 cm) long; margins toothed or entire. RANGE: Texas to Arizona, south to Guatemala. RELATED SPECIES: 2 uncommon Trans-Pecos species. See *V. stenoloba*, below.

95. Skeleton-leaf Goldeneye, *Viguiera stenoloba*
Asteraceae, Sunflower Family

Shrubby perennial 2–4 feet (6–12 dm) tall; stems woody, densely branched, with numerous flowers; in deserts, mountains, dry plains; common. FLOWERS: Spring–fall; yellow heads about 1 inch (2.5 cm) wide, solitary on long stalks; about 12 ray flowers, numerous disk flowers. LEAVES: Alternate or opposite, crowded; blade deeply divided into threadlike, linear or narrowly lanceolate lobes, 1–4 inches (2.5–10 cm) long, ⅟₃₂–³⁄₁₆ inch (1–5 mm) wide; margins entire or few-toothed. RANGE: Texas, New Mexico; Mexico.

96. *Wedelia texana (W. hispida, Zexmania hispida)*
Asteraceae, Sunflower Family

Clump-forming perennial 1–3 feet (3–9 dm) tall; base woody; stems and leaves covered with rough hairs; in sun or partial shade, in dry limestone soil, in fields, and open woods, on hills; abundant.

FLOWERS: Spring–fall; golden yellow-orange heads about 1 inch (2.5 cm) wide; 7–15 ray flowers with notched tips; heads usually solitary on end of long stalk rising well above leaves. LEAVES: Simple, opposite; blade lanceolate, 2–3 inches (5–7.5 cm) long, pointed at both ends; margins with few teeth, occasionally with pair of lobes near base. RANGE: Texas; Mexico. SIMILAR SPECIES: *Jefia brevifolia*, in South Texas and the Trans-Pecos, has leaves usually under 1½ inches (4 cm) long and entire margins.

97. Sleepy Daisy, *Xanthisma texanum*
Asteraceae, Sunflower Family

Taprooted annual 6–36 inches (1.5–10 dm) tall, branching; in sandy plains, along roadsides; common, with 2 varieties. FLOWERS: Spring–fall; lemon-yellow heads, ¾–1½ inches (2–4 cm) wide; ray flowers sharp-pointed; numerous disk flowers, receptacle nearly flat. LEAVES: Simple, alternate; upper leaves small, linear or lanceolate, margins entire; lower leaves to 2½ inches (6.3 cm) long, margins pinnately lobed, merely toothed, or entire. RANGE: Texas, Oklahoma. NOTE: The flower closes up at night, inspiring the common name.

98. Zinnia, *Zinnia grandiflora*
Asteraceae, Sunflower Family

Low shrubby perennial to 8 inches (2 dm) tall; in sunny, dry, calcareous soil, in plains, deserts, hills, mountains; common. FLOWERS: Spring–fall; small yellow heads; 3–6 ray flowers, egg-shaped to round, to ¾ inch (2 cm) long; numerous red or green disk flowers. LEAVES: Simple, opposite; linear to narrowly lanceolate, ½–1 inch (1.2–2.5 cm) long, 1/16 inch (2 mm) wide; margins entire. RANGE: Colorado to Mexico. RELATED SPECIES: *Zinnia acerosa*, in the Trans-Pecos, has white ray flowers; *Z. anomala*, in West Texas, has yellow rays usually under ¼ inch (6 mm) long or lacking, and leaves to ³⁄₁₆ inch (5 mm) wide. Compare with *Psilostrophe* species, page 96.

99. Fringed Puccoon, *Lithospermum incisum*
Boraginaceae, Borage Family

Small, colorful perennial to 1 foot (3 dm) tall; usually in sandy soil, in prairies, fields, edges of woods; widespread and common. FLOWERS: Early spring–summer; bright yellow; 5 petals form narrow tube ½–1⅜ inch (1.2–3.5 cm) long, opening out to flat star ⅜–¾ inch (1–2 cm) in diameter; margins of lobes fringed. LEAVES: Simple, alternate; linear to narrowly oblanceolate, 1–3 inches (2.5–7.5 cm) long (or longer near base); covered with tiny hairs; margins entire. RANGE: Widespread in U.S., Canada, Mexico. NOTE: The seeds are usually produced in buds that develop late in the blooming season and never open. RELATED SPECIES: The fringed petals distinguish this from 9 other Texas species, including two endemics.

100. Bladderpod, *Lesquerella argyraea*
Brassicaceae, Mustard Family

Biennial or perennial, usually with several stems from base, 6–28 inches (1.5–7 dm) tall; in sandy or limestone soil; stems and leaves hairy; common. FLOWERS: Winter–spring; yellow; to ¾ inch (2 cm) wide; 4 petals. LEAVES: Simple, alternate, also basal; stem leaves narrowly linear to broad, to 1½ inches (4 cm) long, with margins wavy, toothed, or entire; basal leaves larger, entire to deeply pinnately lobed. FRUIT: Round or elliptic inflated capsule, resembling a pop bead, ⅛–⅜ inch (4–9 mm) wide; hairless; stalk often S-shaped. RANGE: Texas; Mexico. RELATED SPECIES: 17 other species in Texas, some very similar, 8 endemics, at least 2 endangered. NOTE: The seeds of all species may be used as a peppery seasoning.

101. Plains Wild Indigo, *Baptisia bracteata*
(*B. leucophaea*)
Fabaceae, Legume Family

Low bushy perennial to 2½ feet (8 dm) tall; stems essentially hairless; in sun or partial shade, in sandy woods, fields, prairies; common in some areas, with 2 varieties. FLOWERS: Spring; cream to dark yellow; pea-type, to 1¼ inch (3 cm) long; in drooping or horizontal clusters that may touch the ground. LEAVES: Alternate, compound, with 3 leaflets, leaflets 1–4 inches (2.5–10 cm) long; stipules and bract at base of flower leaflike; margins entire. FRUIT: Inflated pod to 2 inches (5 cm) long, tip tapering. RANGE: Eastern U.S. RELATED SPECIES: This species is the only one with flower stems longer than ⅝ inch (1.5 cm). *Baptisia australis* has blue flowers. *B. alba* has white flowers in erect racemes; *B. nuttalliana*, with hairy stems, has yellow flowers, solitary or few in short racemes. *B. sphaerocarpa* has bright yellow flowers in erect racemes. NOTE: Plants are toxic.

102. Partridge Pea, *Chamaecrista fasciculata*
(*Cassia fasciculata*)
Fabaceae, Legume Family

Erect annual 1–5 feet (3–15 dm) tall; in partial shade, in sandy or alluvial soil, in most open woods and fields, along roadsides; abundant, often in masses; 5 varieties. FLOWERS: Spring–fall; yellow (rarely white); to 1½ (4 cm) inches wide; 5 petals with red dot at base; 1 petal larger, 1 petal curved inward; 10 stamens. LEAVES: Alternate, pinnately compound; 10–30 linear to oblong leaflets to ½ inch (1.2 cm) long; margins entire. FRUIT: Linear pod to 3 inches (8 cm) long. RANGE: Eastern U.S. RELATED SPECIES: Above characters usually distinguish this from 5 other Texas species, 2 of which are prostrate. NOTE: Plants are toxic.

103. *Senna lindheimeriana* (*Cassia lindheimeriana*)
Fabaceae, Legume Family

Bushy perennial to 6 feet (2 m) tall, with velvety stems; in sun, in dry rocky limestone soil, in fields and on hills; common. FLOWERS: Summer–fall; yellow, with red veins in petals; about 1 inch (2.5 cm) wide; 5 petals, not quite equal in size. LEAVES: Alternate, pinnately compound; 8–16 oval to oblong leaflets, ½–2 inches (1–5 cm) long; surface velvety; margins entire. FRUIT: Linear flat pod to 3 inches (8 cm) long. RANGE: Southwestern U.S.; Mexico. RELATED SPECIES: Above characters distinguish this from 13 other species. NOTE: Plants are toxic.

104. Two-leaved Senna, *Senna roemeriana*
(*Cassia roemeriana*)
Fabaceae, Legume Family

Erect perennial to 2 feet (6 dm) tall; in sun, in dry caliche or clay soils, in fields and open woods and on hills; common. FLOWERS: Spring–fall; yellow; to 1¼ inches (3 cm) wide; 5 petals, not quite equal in size. LEAVES: Alternate, compound, with 2 leaflets; leaflets narrowly lanceolate, 1–2½ inches (2.5–6.5 cm) long; hairy (hand lens) but hair not velvety or feltlike. FRUIT: inflated or flattened linear pod to 1½ inches (4 cm) long. RANGE: Mexico to New Mexico. RELATED SPECIES: Above characters distinguish this from 3 other 2-leaved sennas. NOTE: Plants are toxic.

105. Scrambled Eggs, *Corydalis* **species**
Fumariaceae, Fumitory Family

Delicate annuals or biennials, 6–15 inches (15–40 cm) tall, either erect or sprawling; in sandy or rocky areas, prairies, fields and woods, on slopes, along streams and roadsides; some species common, but small size makes plants inconspicuous. FLOWERS: Late winter–spring; yellow; ½–1 inch (1.2–2.5 cm) long; bilateral, with tubular spur; in racemes. LEAVES: Alternate or basal, greatly dissected, segments fernlike. FRUIT: Slender capsule to 1½ inches (4 cm) long. RANGE: Varies with species. RELATED SPECIES: 4 Texas species, difficult to distinguish. Two endemic varieties.

106. Spotted Bee Balm, *Monarda punctata*
Lamiaceae, Mint Family

Aromatic annual or perennial to 3 feet (1 m) tall; stems square in cross section; usually in sandy soil; common, with several varieties. FLOWERS: Spring–summer; yellow (occasionally white or pink), spotted with maroon or brown dots; ½–1 inch (1.2–2.5 cm) long; 2-lipped, with upper lip arching; flowers in interrupted spikes; calyx teeth triangular, not threadlike; whorls of leaflike bracts below flowers; bracts white, yellowish, or pale green (sometimes pink or purple), tip tapering but not long or threadlike. LEAVES: Simple, opposite; blades lanceolate, oblong, or linear, varying in size to 4 inches (1 dm) long; margins toothed to nearly entire. RANGE: Eastern U.S. NOTE: Thymol, an ingredient in cough syrup, is derived from the oil. RELATED SPECIES: Above characters distinguish this from 6 other species. See lemon bee balm, page 42.

107. Golden Flax, *Linum berlandieri*
var. *berlandieri (L. rigidum)*
Linaceae, Flax Family

Slender hairless annual usually less than 1 foot (3 dm) tall; in sand, gravel, or rocky soil, prairies, fields; uncommon, with 4 varieties. FLOWERS: Late winter–summer; yellow to peachy orange; base of petal often brick red or reddish purple, with reddish veins extending up halfway or more; 5 petals ⅜–⅝ inch (10–17 mm) long. LEAVES: Simple, alternate; linear to narrowly lanceolate, to 1¼ inches (3 cm) long; erect, crowded on stem. RANGE: Central U.S. RELATED SPECIES: 20 Texas species. Several yellow species difficult to distinguish. NOTE: Native Americans extracted and wove threads from flax stems. Linen is derived from a cultivated species.

108. Rock-nettle, *Eucnide bartonioides*
 Loasaceae, Stickleaf Family

Clump-forming perennial; in sun, in dry limestone soil, clinging to slopes, cliff faces; uncommon and scattered but can form large colonies. FLOWERS: Nearly year-round; bright yellow; 1–3 inches (2.5–8 cm) wide; showy 5-petaled flower, with many long stamens. LEAVES: Simple, alternate; covered with stiff hairs; blade broad, heart-shaped or nearly round; margins toothed and shallowly lobed. RANGE: Texas; Mexico. NOTE: Central Texas population has smaller flowers than those farther west.

109. Velvetleaf Mallow, *Allowissadula holosericea*
 (Wissadula holosericea)
 Malvaceae, Mallow Family

Branching perennial 2–6 feet (6–20 dm) tall; leaves aromatic; in dry rocky soil; common. FLOWERS: Spring–fall; yellow-orange, ¾–2 inches (2–5 cm) wide; 5 petals, numerous stamens. LEAVES: Simple, alternate; blade broad, heart-shaped, 2–8 inches (5–20 cm) long; covered with soft velvety glandular hairs; margins toothed (rarely nearly entire) and may be shallowly 3-lobed. FRUIT: 5-part dry capsule. RANGE: Texas; Mexico. RELATED SPECIES: *Allowissadula lozanii*, in South Texas, has yellow flowers and lacks glandular hairs. 2 South Texas species with margins entire or teeth inconspicuous, and petals under ⁵⁄₁₆ inch (8 mm) long—*Wissadula amplissima* has yellow flowers; *W. periplocifolia* has lanceolate leaves and white flowers. Some *Abutilon* species are difficult to distinguish.

110. Yellow Lotus, Water Chinquapin,
Nelumbo lutea
Nymphaeaceae, Water-lily Family

Showy aquatic perennial rooted in mud; in ponds, shallow lakes, slow-moving streams; locally common but populations few and scattered. FLOWERS: Summer; creamy yellow, up to 10 inches (2.5 dm) wide, with 20 or more tepals; usually raised above water on long stalk. LEAVES: Circular, 1 to 2 feet (3–6 dm) in diameter; usually raised above water on long stalk attached to center of leaf. FRUIT: Flat-topped woody receptacle 4 inches (1 dm) in diameter, containing numerous acornlike nuts. RANGE: Eastern U.S.; southern Ontario. NOTE: The inside of the tubers is edible when baked but is bitter when raw. RELATED SPECIES: *Nelumbo nucifera*, from the Orient, is the only other lotus in the world. Both species are grown as ornamentals. See water-lilies, page 128.

111. Spatterdock, *Nuphar lutea*
subsp. *advena*
Nymphaeaceae, Water-lily Family

Perennial aquatic or wetland plant rooted in mud; in ponds, shallow lakes, slow-moving streams; common. FLOWERS: Early spring–fall; bright yellow waxy cup-shaped blossom, floating or above water; to 1¾ inch (4.5 cm) wide; 4–6 showy sepals, all yellow or outer ones green; true petals numerous, yellow, reduced to resemble stamens. LEAVES: Oval to almost circular, 1 foot (3 dm) in diameter, with rounded notch at point of attachment to stalk; floating or emerging from water. RANGE: Eastern North America; northern Mexico, Cuba. RELATED SPECIES: See water-lilies page 128.

112. Sundrops, Evening Primrose,
Calylophus berlandieri (C. drummondianus)
Onagraceae, Evening Primrose Family

Erect to spreading perennial to 4–32 inches (1–8 dm) tall; in sun, in sandy or rocky soil, in fields and on hills. FLOWERS: Spring–summer; 1–2 inches (2.5–5 cm) wide; 4 yellow petals, center and stigma black or yellow, with stigma extending above stamens; bud 4-angled, sepals with keeled midribs. LEAVES: Alternate, simple; narrow to linear, 1–3 inches (2.5–7.5 cm) long; margins entire or toothed. RANGE: Kansas and Colorado to Mexico. RELATED SPECIES: Keeled bud and protruding stigma distinguish this species from 4 other species in Texas. The single knobby stigma of *Calylophus* distinguishes the genus from *Oenothera*.

113. Water Primrose, *Ludwigia uruguayensis*
Onagraceae, Evening Primrose Family

Erect or ascending wetland perennial to 3 feet (1 m) tall; softly hairy; floating in or growing along streams, ditches, ponds; uncommon. FLOWERS: Summer; yellow; 1–2 inches (2.5–5 cm) wide; 5 petals, 10 stamens. LEAVES: Simple, alternate; blade oblanceolate or elliptic, to 4 inches (1 dm) long; margins entire. RANGE: Southeastern U.S., through South America. RELATED SPECIES: 9 species in Texas, some very similar; some with opposite leaves or 4 petals.

114. Fluttermill, *Oenothera macrocarpa*
(O. missouriensis)
Onagraceae, Evening Primrose Family

Low-growing perennial with spreading stems to 2 feet (6 dm) long or with no stems; on sunny, dry limestone hills, prairies;

uncommon, with 2 subspecies. FLOWERS: Spring–summer; bright yellow (fading reddish); 4 lobes 1–2 inches (2.5–5 cm) long; thin tube 2–6 inches (5–15 cm) long below lobes; stigma cross-shaped; 8 stamens. LEAVES: Simple, alternate or basal; narrowly to broadly lanceolate or elliptic, to 4 inches (1 dm) long; margins entire or with few widely spaced teeth. FRUIT: 4-winged capsule 1–2½ inches (2.5–6.5 cm) long, wings ¼–¾ inch (7–20 mm) wide. RANGE: South-central U.S. RELATED SPECIES: Above characters distinguish this from 25 other species. *Oenothera brachycarpa*, in the Trans-Pecos, is very similar; its leaves can be deeply lobed and the wings on fruit are less than ³⁄₁₆ inch (5 mm) wide. See prairie primrose, page 49.

115. Yellow Wood Sorrel, *Oxalis stricta (O. dillenii)*
Oxalidaceae, Wood Sorrel Family

Small clump-forming perennial usually under 1 foot (3 dm) tall; in wide variety of habitats, woods, fields, lawns, gardens; common. FLOWERS: Late winter–fall; yellow; often much smaller than 1 inch (2.5 cm) wide; 5 petals, each ¼–½ inch (5–12 mm) long. LEAVES: Alternate or clustered; palmately compound, with 3 heart-shaped leaflets ⅜–1¼ inch (1–3 cm) wide; hairy. FRUIT: Slender, erect ridged capsule; brown seeds with white marks on ridges (hand lens). RANGE: Eastern and central U.S.; introduced into Europe. NOTE: The leaves and fruit can be added to salads but should not be eaten in large quantity. Oxalic acid provides the tart flavor. RELATED SPECIES: Above characters distinguish this from 6 other yellow Texas species. See purple wood sorrel, page 49.

116. Yellow Prickly Poppy, Cardo Santo,
Argemone mexicana
Papaveraceae, Poppy Family

Erect prickly annual to 4 feet (1.2 m) tall; sap yellow; in full sun, in disturbed soil, in fields, along roadsides. FLOWERS: Spring;

bright or pale yellow; 1½–3 inches (4-7.5 cm) wide; 6 petals, 1–1½ inches (2.5–4 cm) long; 20–75 yellow stamens; buds prickly. LEAVES: Simple, alternate, also basal; stems leaves to 3 inches (8 cm) long; basal leaves to 7 inches (1.8 dm); margins pinnately lobed, armed with stiff prickles. FRUIT: Elliptic prickly capsule filled with many tiny seeds; when capsule dries, small holes at top open up, resembling a pepper shaker. RANGE: Eastern U.S. to South America. RELATED SPECIES: *Argemone aenea* has 150–250 reddish stamens and slightly larger pale yellow to bronze flowers; 6 other species, some white or rose. See other prickly poppies, pages 50 and 129. NOTE: Sap is a skin irritant.

117. Large Buttercup, *Ranunculus macranthus*
Ranunculaceae, Buttercup Family

Perennial with erect or sprawling stems 6–36 inches (1.5–10 dm) long; covered with coarse hairs; in wet soil, in ditches, swamps, and woods, along streams and seeps; common. FLOWERS: Spring–summer; yellow; ½–1½ inches (1.2–4 cm) wide; 8–20 glossy petals, numerous stamens. LEAVES: Alternate, also basal; most leaves compound, with 3–7 leaflets; leaflets deeply lobed to entire, highly variable. RANGE: Southwestern U.S.; Mexico. NOTE: The hairs and sap can cause skin irritation. RELATED SPECIES: This is the most widespread of 19 Texas species. Others have smaller flowers; some have white flowers.

118. Pitcher Plant, *Sarracenia alata*
Sarraceniaceae, Pitcher Plant Family

Insectivorous perennial in acid bogs, pine woods; becoming rare. FLOWERS: Early spring–summer; greenish yellow, showy; 1½–2½ inches (4–6.5 cm) long; 5 petals, 5 sepals, 3 bracts; bloom nodding on end of leafless stalk. LEAVES: Basal; tubular trumpets to 28 inches (7 dm) tall; greenish yellow, may have reddish veins; capped with hood. RANGE: Alabama to Texas. NOTE: Acid bogs are low in minerals, so these plants extract nitrogen from insects. The prey is trapped and digested in the liquid held in the leaves.

119. Indian Paintbrush, *Castilleja purpurea*
   var. *lindheimeri*
   Scrophulariaceae, Figwort Family

Erect perennial 8–18 inches (2–4.5 dm) tall; in full sun, in dry caliche or sandy soil, hills, in prairies; uncommon. FLOWERS: Spring; showy, reddish orange to yellow-orange bracts enclose inconspicuous flowers; flowers tubular, creamy or greenish, may be tinged with color, to 1½ inches (4 cm) long; bracts and flowers form dense spike. LEAVES: Simple, alternate; 1–3 inches (2.5–7.5 cm) long; margins with deep linear or lanceolate lobes. RANGE: Endemic to Texas. RELATED SPECIES: *Castilleja purpurea* var. *purpurea* has pink to lavender flowers; *C. p.* var. *citrina* has yellow flowers; 8 other Texas species, with various flower colors. See *Castilleja indivisa*, page 54.

120. Mullein, *Verbascum thapsus*
   Scrophulariaceae, Figwort Family

Densely hairy biennial; first year produces basal rosette of large leaves; second year, single flowering stalk to 6 feet (2 m) tall; in sun, in disturbed ground, fields, and open woods, along roadsides; common. FLOWERS: Spring–fall; yellow; about 1 inch (2.5) wide; in spike 6–24 inches (1.5–6 dm) long at top of tall stalk. LEAVES: Simple, alternate or basal; broad, to 16 inches (4 dm) long at base, upper leaves smaller; surface feltlike; margins entire or toothed. RANGE: Native of Europe, now widely naturalized in North

America and throughout Texas. NOTE: The leaves provide a yellow dye for wool and a tea to soothe sore throats. Also fondly known as the toilet paper of the woods. RELATED SPECIES: 2 East Texas species lack the woolly leaves.

121. Tomatillo, Ground-cherry,
Physalis cinerascens (P. viscosa)
Solanaceae, Nightshade Family

Hairy perennial to 10 inches (2.5 dm) tall; in woods and fields or along beaches; common, with 3 varieties. FLOWERS: Spring–fall; yellow, with dark spots in center; 5 petals united in blossom about ¾ inch (2 cm) wide; stamens protruding. LEAVES: Simple, alternate or in pairs; blades egg-shaped, triangular, or elliptic; hairs stellate; margins toothed or wavy. FRUIT: Miniature yellow tomato-like berry to ½ inch (1.2 cm) in diameter, completely enclosed by inflated husk; husk is ribbed, to 1 inch (2.5 cm) long. RANGE: Throughout Texas; south-central U.S.; Mexico. NOTE: The ripe fruit is edible, but unripe fruit and foliage are toxic. RELATED SPECIES: 13 other species, some very similar, two endemic.

122. Buffalo Bur, *Solanum rostratum*
Solanaceae, Nightshade Family

Branching annual 1–3 feet (3–10 dm) tall; stems and leaves covered with yellowish prickles; in sun, in disturbed soils; common and weedy. FLOWERS: Spring–fall; yellow; 1 inch (2.5 cm) wide; 5 petals united. LEAVES: Alternate, 2–8 inches (5–20 cm) long; deeply pinnately divided. FRUIT: Burlike dry capsule about ½ inch (1.2 cm) in diameter, covered with prickles; poisonous. RANGE: Throughout Texas, north to Nebraska, spreading outside natural range. RELATED SPECIES: The yellow flowers distinguish this from several other species with prickly fruit. See *S. elaeagnifolium*, p. 87.

WHITE OR GREEN FLOWERS

---

123. Water-willow, *Justicia americana*
Acanthaceae, Acanthus Family

Aquatic perennial to 3 feet (1 m) tall; in stream beds, shallow water, mud; common. FLOWERS: Spring–fall; white to violet with purple dots on interior; short tube to ½ inch (1.2 cm) long; with 5 lobes that may curl slightly; flowers in tight head or short spike, at top of long stalk. LEAVES: Simple, opposite; linear to lanceolate or oblanceolate, 2–6 inches (5–15 cm) long; margins entire. RANGE: Eastern U.S.; Canada. RELATED SPECIES: One of the most common of 10 species, 2 of which are endangered— *Justicia runyonii* in South Texas, and *J. wrightii* endemic to the Trans-Pecos and western Edwards Plateau.

124. Arrowhead, Flecha de Agua,
Sagittaria longiloba
Alismataceae, Water Plantain Family

Aquatic plant with leaves above or below water; in shallow water, ditches, ponds, swamps; common in South Texas. FLOWERS: Spring–fall; white, to 1½ inches (4 cm) wide; 3 petals; flowers in whorls of 3 on tall stalk. LEAVES: Simple, all basal; blade triangular, arrow-shaped, with narrow, linear, or lanceolate basal lobes, lobes longer than main body of blade. RANGE: Southwestern U.S.; Mexico. RELATED SPECIES: Above characters distinguish this from 8 other species. Some species have leaves with broad lobes; others have unlobed leaves. NOTE: The tiny tubers at the ends of the underwater roots of all species are edible. Cook them like potatoes.

125. Rain Lily, Cebollita, *Cooperia drummondii*
Amaryllidaceae, Amaryllis Family

Slender perennial 6–18 inches (1.5–4.5 dm) tall, emerging from bulb, flower solitary; often appearing after rain, in lawns, fields, prairies, open woods; common to abundant. FLOWERS: Late summer–fall, occasionally spring; 6 white tepals (may be pink-tinged), to 1 inch (2.5 cm) long; floral tube 3–7 inches (8–18 cm) long from base of tepal lobes to swollen ovary. LEAVES: Basal; linear, grasslike. RANGE: South-central U.S.; Mexico. RELATED SPECIES: *Cooperia pedunculata*, a spring and summer bloomer, has a floral tube to 1½ inches (4 cm) long; *C. traubii*, a delicate species, is endemic to the northern coast. Other species have yellow flowers. *C. smallii* is endemic to Cameron County. NOTE: Plants are toxic.

126. Spider Lily, *Hymenocallis liriosme*
Amaryllidaceae, Amaryllis Family

Fleshy perennial to 3 feet (1 m) tall, from bulb; in wetlands and ditches, along streams; abundant. FLOWERS: Spring, rarely to summer; fragrant; white; to 7 inches (1.8 dm) wide; 6 long, narrow tepals usually under ¼ inch (5 mm) wide surround cup-shaped structure (crown); stamens attached to crown. LEAVES: Basal; slender, grasslike, to 30 inches (7.5 dm) long, usually less than ¾ inch (2 cm) wide. RANGE: South-central U.S. NOTE: The plant is toxic. RELATED SPECIES: *Hymenocallis caroliniana*, in South Central and East Texas, has slightly larger flowers and wider leaves; *H. eulae*, in Southeast Texas, blooms in summer and also has wider leaves (may be a variety of *H. liriosme*).

127. Water Hemlock, Musquash Root,
Cicuta maculata
Apiaceae, Carrot Family

Attractive but poisonous perennial 2–8 feet (6–25 dm) tall; stem may be purple-spotted or -streaked; in wetlands, along streams and lakes in marshes; uncommon. FLOWERS: Spring–fall; white or greenish; minute flowers in compound umbel 2–8 inches (5–20 cm) wide. LEAVES: Alternate, pinnately compound (twice or thrice compound); 1–3 feet (3–10 dm) long; base usually sheathing the stem; leaflets lanceolate, 1–3 inches (2.5–7.5 cm) long; margins toothed. RANGE: Eastern U.S.; Canada. NOTE: This is possibly the most deadly poisonous wild flowering plant in the U.S. RELATED SPECIES: Water hemlock is easily confused with nonpoisonous members of the carrot family; thus we discourage the use of any wild carrots as food. The following characters help identify water hemlock (one plant may have any or all of these characters): The root usually forms a cluster of tubers rather than a single taproot; when cut in half, the pith of the swollen section of stem just above the root contains a series of horizontal layers, forming small chambers; a yellow pleasant-smelling oil may seep out of this pith. *Cicuta mexicana,* in East Texas and along the coast, is very similar and may be a variety of *C. maculata.*

128. Poison hemlock, *Conium maculatum*
Apiaceae, Carrot Family

Poisonous biennial weed with hairless, often purple-spotted or -streaked stem 2–9 feet (6–30 dm) tall; first year produces basal rosette of large leaves 1–2½ feet (3–8 dm) long; second year, tall flower stalk; in wetlands, along streams; locally abundant. FLOWERS: Spring–summer; white; minute flowers in compound umbel 2–6 inches (5–15 cm) wide. LEAVES: Alternate, pinnately compound (twice or thrice compound); base usually sheathing the stem; leaflets greatly dissected, like carrot leaf; disagreeable musky odor. RANGE:

Introduced from Eurasia; widely naturalized. NOTE: This is one of the most deadly poisonous wild flowering plants; Socrates was put to death with a cup of hemlock tea. RELATED SPECIES: The leaves and taproot are easily confused with those of wild carrot (below), thus we discourage the use of wild carrot as food.

129. Queen Anne's Lace, Wild Carrot, *Daucus carota*
Apiaceae, Carrot Family

Taprooted biennial to 4 feet (1.2 m) tall; stem rough; hairy; or smooth; along roadsides, in fields and disturbed ground; scattered in Texas, abundant in eastern U.S. FLOWERS: Spring–summer; tiny white flowers (rarely pink or yellow) in large flat to rounded compound umbel; center flower purple or red; cluster forms cup as seeds mature. LEAVES: Alternate, pinnately compound; dissected into narrow fernlike segments. RANGE: Introduced from Eurasia, widely naturalized in U.S., mainly in Northeast and South Texas. NOTE: This is considered a wild variety of the garden carrot. RELATED SPECIES: Compare with the toxic water hemlock and poison hemlock, above. *Daucus pusillus*, throughout Texas, has small compact umbels, that lack a red central flower.

130. Beggar's Ticks, *Torilis arvensis*
Apiaceae, Carrot Family

Taprooted annual 6–24 inches (1.5–6 dm) tall; stems branching, covered with rough hairs; winter plant a basal rosette of leaves; in moist, sunny locations, in disturbed soil, in fields and lawns; abundant and weedy. FLOWERS: Spring–summer; white; minute flowers in compound umbels ½–1½ inches (1.2–4 cm) wide on stalks ½–5 inches (1.2–12 cm) long. LEAVES: Alternate, much pinnately divided; 2–6 inches (5–15 cm) long; segments egg-shaped or lanceolate, deeply cut, parsleylike. FRUIT: Tiny, seed-like, coated with hooked bristles. RANGE: Native of Old World, now widely naturalized. NOTE: In summer when walking through overgrown fields, you are sure to come home with socks full of the fruits. RELATED SPECIES: *Torilis nodosa* has flowers in compact heads with short or no stalks, opposite the leaf axils. Compare with other members of the family.

131. Green Milkweed, *Asclepias asperula*
Asclepiadaceae, Milkweed Family

Spreading to erect perennial to 2 feet (6 dm) tall; sap milky; in full sun, in sandy or rocky soil, along roadsides, in fields, open woods; common. FLOWERS: Spring–fall; 5 greenish-yellow cupped petals, with 5 horn-shaped reddish or greenish appendages (hoods); many flowers in rounded umbel. LEAVES: Simple, opposite or alternate; lanceolate to almost linear, 3–8 inches (7.5–20 cm) long, ¼–1¼ inches (6–30 mm) wide, often folded lengthwise; surface often rough; margins entire. FRUIT: Pod to 5 inches (1.3 dm) long; flat seeds attached to fluffy hairs. RANGE: Texas. RELATED SPECIES: Above characters distinguish this from 35 other species. NOTE: Milkweeds are poisonous.

132. Yarrow, Milfoil, *Achillea millefolium*
Asteraceae, Sunflower Family

Erect perennial to 3 feet (1 m) tall; covered with soft, woolly hairs; often in partial shade, in various soils, in fields, at edges of woods; common. FLOWERS: Spring–summer; white (occasionally pink); tiny heads about ¼ inch (6 mm) wide, ray flowers white, disk flowers white or yellow; heads form dense round-topped clusters at top of stem. LEAVES: Alternate; pinnately dissected into many threadlike sections; soft, grayish, resembling feathers or fern leaves; aromatic. RANGE: Native of Europe and Asia, now growing wild in much of U.S. NOTE: The leaves have medicinal value but also contain toxic compounds. Flowers make a nice yellow dye. SIMILAR SPECIES: Yarrow may be confused with wild carrot, page 117.

133. Lazy Daisy, *Aphanostephus skirrhobasis*
Asteraceae, Sunflower Family

Taprooted hairy annual 4–20 inches (1–5 dm) tall; one or numerous leafy stems spread out from base; in sun; in sandy soil; common, with 3 varieties. FLOWERS: Spring–summer; heads ½–1½ inches (1.2–4 cm) wide; ray flowers white to lavender or rose, often with lavender stripe on bottom; disk flowers yellow. LEAVES: Simple, alternate, also basal; basal leaves about 3 inches (8 cm) long, often deeply lobed though may be merely toothed or entire, gradually becoming smaller and less divided to entire up the stem. RANGE: Southeastern U.S.; Mexico. RELATED SPECIES: 3 other species in Texas, difficult to distinguish.

134. Aster, *Symphyotrichum (Aster)* **species**
Asteraceae, Sunflower Family

Annuals or perennials; common. FLOWERS: Fall; some species also in late summer or winter; heads small, rays white, pink, blue, or purple (not yellow); disk flowers usually yellow. LEAVES: Simple, alternate; usually lacking petioles; margins not deeply cut. RANGE: Varies with species. RELATED SPECIES: 20 species in Texas, often difficult to distinguish. See Texas aster, p. 28. Of 3 endemic species, 2 may be endangered.

135. Fleabane, *Erigeron* **species**
Asteraceae, Sunflower Family

Slender annuals or perennials mostly under 1 foot (4 dm) tall; often hairy. FLOWERS: Most species bloom in spring; heads, about ½–1 inch (1.2–2.5 cm) wide, white (occasionally pale lavender) with yellow center; ray flowers numerous, linear or threadlike, usually in 2 rows; many disk flowers. LEAVES: Alternate. RELATED SPECIES: 15 species in Texas, difficult to distinguish. Also resemble several other genera. *E. mimegletes*, endemic to western Edwards Plateau and southern Trans-Pecos, may be endangered.

136. Blackfoot Daisy,
Melampodium leucanthum
Asteraceae, Sunflower Family

Perennial to 1½ feet (4.5 dm) tall, often with many branches spreading out from base; on dry limestone soil; common. FLOWERS: Nearly year-round; heads to 1½ inch (4 cm) wide, white with yellow center; 8–10 white ray flowers, many yellow disk flowers; dark lines on underside of rays. LEAVES: Simple, opposite; narrow, linear, oblong, or oblanceolate, under 2 inches (5 cm) long; hairs rough; margins entire or pinnately lobed. RANGE: Southwestern U.S.; Mexico. RELATED SPECIES: *Melampodium cinereum*, in South Texas, is very similar; *M. strigosum*, in the Trans-Pecos, has inconspicuous yellow rays.

137. Compass plant, White Rosinweed,
Silphium albiflorum
Asteraceae, Sunflower Family

Perennial, 1–3 feet (3–10 dm) tall; foliage and stem very rough, sticky; in dry limestone soil, along roadsides, in prairies; uncommon. FLOWERS: Spring–summer; white heads 1½–3 inches (4–7.5 cm) wide; numerous ray and disk flowers. LEAVES: Alternate; deeply pinnately divided into numerous lobes; 4–12 inches (1–3 dm) long; stiff. RANGE: Endemic to Central Texas. RELATED SPECIES: *Silphium laciniatum* is similar but has yellow flowers; 3 other yellow species without deeply lobed leaves.

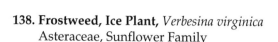

138. Frostweed, Ice Plant, *Verbesina virginica*
Asteraceae, Sunflower Family

Erect perennial, 1–12 feet (0.3–3.6 m) tall; leaf bases form wings on stem; usually in moist soil, along streams, in shade of trees; abundant. FLOWERS: Late summer–fall; white to greenish white; heads ¼–½ inch (6–12 mm) wide; 3–4 ray flowers (never more than 7), less than 15 disk flowers; in large terminal clusters. LEAVES: Simple, alternate; blade broad, lanceolate to egg-shaped, 4–10 inches (1–2.5 dm) long; surfaces hairy; margins toothed or wavy. RANGE: Eastern U.S., north to Pennsylvania. NOTE: In the first freeze, the sap ruptures the stems and forms a delicate curtain of ice crystals. RELATED SPECIES: 1 other white-flowering species, *Verbesina microptera*, grows south of San Antonio and has 9 or more ray flowers and more than 20 disks per head; 6 other species have yellow flowers.

139. Mayapple, Mandrake, *Podophyllum peltatum*
Berberidaceae, Barberry Family

Perennial to 2 feet (6 dm) tall; in lush forests, moist shady fields; locally abundant, in large colonies. FLOWERS: early spring; fragrant; creamy white to rose; cup-shaped, to 3 inches (8 cm) wide, solitary and hidden, hanging from axil of 2 leaves; 6 or 9 petals. LEAVES: 1 or 2 per plant, on long stems, resembling umbrellas; to 1 foot (3 dm) or more in diameter; margins toothed and deeply palmately lobed. FRUIT: Golden yellow or purplish oval berry 2 inches (5 cm) long. RANGE: Eastern U.S. and Canada. NOTE: Though the ripe fruit is edible, the unripe fruit and all other parts are highly poisonous.

140. Shepherd's Purse, Panquecillo, *Capsella bursa-pastoris*
Brassicaceae, Mustard Family

Small annual to 2 feet (6 dm) tall; weed in lawns and fields and along roadsides; widespread and common. FLOWERS: winter–spring; white; minute; 4-petaled. LEAVES: Basal in rosette, deeply pinnately cut into lobes; stem leaves toothed or entire. FRUIT: Flattened triangular to heart-shaped capsule, known as a silique, to ⅜ inch (1 cm) wide, resembling a tiny purse. RANGE: Worldwide, introduced and naturalized throughout Texas. NOTE: The capsules and young, tender leaves may be used as a peppery spice. The scientific specific name means "shepherd's purse."

141. Peppergrass, Lentejilla, *Lepidium virginicum*
Brassicaceae, Mustard Family

Annual or biennial much-branched herb to 2 feet (6 dm) tall; in disturbed soils, in open areas, fields, and lawns, along roadsides; widespread and abundant. FLOWERS: Winter–summer; white; minute; 4-petaled; dense along upper part of stem. LEAVES: Simple, alternate, also basal; upper stem leaves narrow, lanceolate to linear, margins toothed or entire; basal and lower stem leaves deeply pinnately lobed. FRUIT: Nearly circular, flattened capsule, known as a silicle, ⅛ inch (4 mm) long. RANGE: Native of Europe, naturalized throughout Texas and U.S. RELATED SPECIES: 7 other species in Texas, some difficult to distinguish. NOTE: The capsules and young leaves of all species can be used as a peppery seasoning.

142. Clammyweed, *Polanisia dodecandra*
Capparaceae, Caper Family

Sticky hairy annual 1–3 feet (3–10 dm) tall; foliage with unpleasant aroma; in variety of habitats, usually in open areas; common, with 3 subspecies. FLOWERS: Spring–fall; white or rose to purple; 4 petals usually ¼–⅝ inch (6–16 mm) long, of unequal sizes, with narrow base, notched tip; pink to purple stamens, longer than petals; flowers in clusters. LEAVES: Alternate, palmately compound, with 3 leaflets (upper leaves may be simple); leaflets broad, ½–2 inches (1.2–5 cm) long; margins entire. FRUIT: Slender erect pod to 3 inches (7.5 cm) long. RANGE: Throughout Texas; Mexico through central U.S. to Canada. RELATED SPECIES: 3 other species in Texas. *Cleome* species are similar but have pods held horizontal or drooping.

143. Chickweed, *Stellaria media*
Caryophyllaceae, Pink Family

Low mat-forming annual or perennial; in moist disturbed soil, in lawns, woods; common and weedy. FLOWERS: Winter–spring; white; ¼ inch (6 mm) wide; 5 petals deeply cleft, so appearing to have 10 petals. LEAVES: Simple, opposite; blade egg-shaped or elliptic, ½–1½ inches (1.2–4 cm) long; upper leaves often sessile, lower leaves with petiole; margins entire. RANGE: Introduced from Eurasia, now widely naturalized in eastern half of Texas. RELATED SPECIES: 2 less common species, with leaves heart- or egg-shaped or triangular and upper leaves on long petioles—*Stellaria cuspidata*, in mountains of the Trans-Pecos, and *S. prostrata*, in South and Central Texas. NOTE: Greens are edible, high in Vitamin C.

144. Snow-on-the-prairie, *Euphorbia bicolor*
Euphorbiaceae, Spurge Family

Hairy annual to 3 feet (1 m) tall; in clay, in open areas, fields; abundant. FLOWERS: Summer–fall; inconspicuous tiny green flowers (many male flowers and 1 female) enclosed by 5 small white petallike structures. LEAVES: Simple, alternate, with upper leaves often in whorls; showy leaflike bracts with white margins, clustered below flowers; bracts narrow, 1–4 inches (2.5–10 cm) long, ¼–½ inch (6–12 mm) wide; stem leaves slightly broader, elliptic, rarely with white margins; surfaces usually densely hairy; margins entire. RANGE: Texas, Oklahoma, Arkansas, and Louisiana. RELATED SPECIES: Snow-on-the-mountain, (*E. marginata*), in the western half of Texas, has bicolored bracts usually about 1 inch long (2.5 cm), the leaves are broadly lanceolate to egg-shaped, and are nearly hairless; the upper leaves may have white margins. Three endemics. NOTE: The caustic milky sap is toxic, the sap and hairs can irritate the skin.

145. Bull Nettle, Mala Mujer, *Cnidoscolus texanus*
Euphorbiaceae, Spurge Family

Bushy perennial 1–3 feet (3–9 dm) tall, covered with stinging hairs; milky sap; in sand and various other soils, in fields, disturbed areas, openings in woods; widespread and common. FLOWERS: Spring–fall; fragrant; white; 5-lobed (sometimes 4-lobed) trumpet about 1½ inches (4 cm) wide. LEAVES: Simple, alternate; palmately lobed blade 2–6 inches (5–15 cm) long; covered with stinging hairs. FRUIT: Oblong edible nuts encased in 3-parted capsule coated with stinging hairs. RANGE: Arkansas and Oklahoma to Mexico. NOTE: Some people have a severe allergic reaction to contact with the stinging hairs. SIMILAR SPECIES: The true stinging nettles (*Urtica* species) have stinging hairs but do not have showy flowers.

146. White Clover, *Trifolium repens*
Fabaceae, Legume Family

Low mat-forming perennial 2–8 inches (0.5–2 dm) tall; with crawling stems that root at the nodes; in moist, sunny locations, invades lawns, roadsides; common and weedy. FLOWERS: Early spring–summer; white (sometimes pinkish); tiny tubular pea-type flowers on short stems; flowers in dense spherical heads about 1 inch (2.5 cm) wide; head solitary on erect leafless stalk, held above leaves. LEAVES: Alternate, palmately compound with 3 oval or circular leaflets ½–1 inch (1.2–2.5 cm) long, often with light green band circling base; margins with tiny teeth. RANGE: Native of Europe, now widely naturalized. RELATED SPECIES: Above characters distinguish this from 13 other species. *Trifolium pratense*, very similar, has pink to red flowers (rarely white). See crimson clover, page 35.

147. Wild Geranium, Cranesbill,
  Geranium carolinianum
  Geraniaceae, Geranium Family

   Low-growing annual or biennial; stem usually spreading and less than 1 foot (3 dm) tall; in lawns, fields, dry woods; common. FLOWERS: Spring; white to pink; about ½ inch (1.2 cm) wide; 5 petals. LEAVES: Simple, alternate, or whorled below flower, also basal; rounded outline, 1–3 inches (2.5–7.5 cm) wide; deeply palmately lobed; often red tinged. FRUIT: Seeds attached to erect beak to ¾ inch (2 cm) long, resembling a crane's bill; beak springs open to eject seeds. RANGE: Most of eastern and midwestern U.S. into Canada. RELATED SPECIES: This is the most widespread of 5 Texas species. Compare with Texas storksbill, page 38.

148. Horehound, Marrubio, *Marrubium vulgare*
  Lamiaceae, Mint Family

   Aromatic perennial usually under 3 feet (1 m) tall; stems woolly, square in cross section; in disturbed soil, along roadsides, in fields, farms, even mountains; abundant. FLOWERS: Nearly year-round; tiny white flowers in tight whorls at leaf bases; 2-lipped, bilateral. LEAVES: Simple, opposite; blade egg-shaped to nearly round or heart-shaped, 1–2 inches (2.5–5 cm) long; wrinkled-looking surface;

margins toothed and wavy. RANGE: Native of Eurasia, now naturalized worldwide; throughout Texas. NOTE: This bitter herb has been used to make horehound tea for treating sore throats.

149. Crow Poison, False Garlic, *Nothoscordum bivalve*
Liliaceae, Lily Family

Slender perennial 4–20 inches (1–5 dm) tall, from bulb; in sun or partial shade, in lawns, fields, prairies, open woodlands; abundant, often in large colonies. FLOWERS: Year-round, most common in spring and fall; fragrant; creamy white; 6 tepals to ½ inch (1.2 cm) long, with a dark stripe on outside; several flowers in umbel at top of leafless stem; individual flower stalks ¾–2 inches (2–5 cm) long or more. LEAVES: Basal; linear, grasslike, ⅟₁₆–⁵⁄₁₆ inch (2–8 mm) wide; some as tall as flower stalk. RANGE: Throughout Texas; eastern U.S., north to Virginia; Mexico. NOTE: This toxic plant can be confused with wild onions, (see page 45) but has a musky, not oniony, odor.

150. Death Camas, *Zigadenus nuttallii*
Liliaceae, Lily Family

Perennial 1–2½ feet (3–7.5 dm) tall, from bulb; in calcareous, rocky soil, in prairies, hills; uncommon. FLOWERS: Spring; white or creamy yellow; 6 tepals to ⁵⁄₁₆ inch (8 mm) long; flowers on stems ½–1¼ inches (1.2–3 cm) long; in dense many-flowered racemes (rarely branching panicles). LEAVES: Alternate, but mostly basal; linear, grasslike from base, 6–24 inches (1.5–6 dm) long, to ½ inch (1.2 cm) wide; stem leaves much reduced. RANGE: Southeastern and south-central U.S. RELATED SPECIES: 4 other species, more restricted in range or rare. Compare with other members of the family. NOTE: All parts of

Zigadenus species are deadly poisonous and have caused death of livestock.

151. White Stickleaf, *Mentzelia nuda*
Loasaceae, Stickleaf Family

Perennial to 4 feet (1.2 m) tall; stem often white; usually in sandy soil, in fields, prairies; common in some areas. FLOWERS: Spring–fall; white (sometimes pale yellow or greenish); 10 (sometimes 12) petals 1–1½ inches (2.5–4 cm) long; many stamens. LEAVES: Simple, alternate; linear to lanceolate; surface sandpapery rough; sessile but leaf base narrow; margins with numerous large teeth. FRUIT: Cylindrical capsule to 1¼ inches (3 cm) long; seeds winged. RANGE: Central U.S. RELATED SPECIES: 14 other species, most with yellow flowers, one endemic.

152. Devil's Claw, Uña de Gato,
Proboscidea louisianica
Martyniaceae, Unicorn Plant Family

Bushy annual 1–3 feet (3–10 dm) tall, covered with sticky hairs; often in sand, in meadows, disturbed ground, and fields, along riverbanks, common in some areas. FLOWERS: Summer–fall; pink to white, dotted with purple or brown, throat yellow; 5-lobed trumpets 1–2 inches (2.5–5 cm) long; in a loose raceme with 8–20 flowers, a few opening at a time. LEAVES: Simple, mostly opposite; blade heart-shaped to round, 2–12 inches (5–30 cm) wide; margins entire, wavy. FRUIT: Fleshy hooked pod; when ripe, splits open to form 2 woody claws. RANGE: Southern U.S.; also in cultivation. NOTE: Papago Indians weave the claws into baskets. The very young pods are edible when cooked. RELATED SPECIES: Above characters distinguish this from 5 other species, 2 of which are endangered.

153. White Water-lily, Ninfa Acuática, *Nymphaea odorata*
Nymphaeaceae, Water-lily Family

Large aquatic perennial rooted in mud; in ponds, shallow lakes, slow-moving streams; uncommon. FLOWERS: Spring–fall; fragrant; white to pink; 2–6 inches (5–15 cm) in diameter, usually floating; numerous harp-pointed petals (usually 25 or more), many stamens; 4 petallike sepals, may be purple-tinged. LEAVES: Nearly circular, to 10 inches (2.5 dm) wide, with pie-shaped wedge at point of attachment to stalk; usually floating; bottom may be reddish. RANGE: Eastern U.S. and Canada. RELATED SPECIES: *Nymphaea elegans* has blue to pale violet flowers; *N. mexicana* has yellow flowers. See spatterdock, page 108, and yellow lotus, page 108. NOTE: Water-lilies are often grown as ornamentals.

154. Ladies' Tresses, *Spiranthes cernua*
Orchidaceae, Orchid Family

Erect single-stemmed perennial 6–16 inches (1.5–4 dm) tall; in moist or wet soils of woods, swamps, and ditches, also in dry rocky areas along streams or seeps; rare. FLOWERS: Late summer–fall; fragrant; white; delicate, ¼–½ inch (6–12 mm) long; in spiraling rows on upper stalk. LEAVES: Simple, basal; elliptic or egg-shaped; stem leaves reduced, lanceolate. RANGE: Eastern and central U.S.; Canada. RELATED SPECIES: 13 other species, most in East Texas or the Trans-Pecos, some very similar. One endemic species, *S. parksii*, is endangered. NOTE: Because of a symbiotic relationship with soil fungi, orchids seldom survive transplanting.

155. White Prickly Poppy,
Argemone albiflora subsp. *texana*
Papaverceae, Poppy Family

Erect prickly annual or biennial to 3 feet (1 m) or taller; sap yellow; in full sun, in disturbed soil, in fields, along roadsides; common. FLOWERS: Spring–fall; white; to 4 inches (1 dm) wide; 6 delicate, wrinkled petals; numerous yellow or red stamens; buds prickly. LEAVES: Simple, alternate, also basal; blade to 9 inches (2.2 dm) long; gray-green surface may be mottled; margins pinnately lobed, armed with stiff prickles. FRUIT: Elliptic prickly capsule filled with many tiny seeds; when capsule dries, small holes at top open up, resembling a pepper shaker. RANGE: South-central U.S. RELATED SPECIES: 7 other species, some yellow or rose. See other prickly poppies, pages 50 and 110. NOTE: Sap is a skin irritant.

156. Small Pokeweed, Coralito, *Rivina humilis*
Phytolaccaceae, Pokeweed Family

Erect to sprawling perennial usually under 1 foot (3 dm), but occasionally to 5 feet (1.5 m) tall; in shade, in woods or shrubbery, along streams, and on limestone; common. FLOWERS: Spring–fall; white, greenish, or pinkish red; tiny; 4 tepals; in racemes. LEAVES: Simple, alternate; blade egg-shaped to lanceolate, usually 1–3 inches (2.5–7.5 cm), sometimes to 6 inches (1.5 dm) long; margins wavy. FRUIT: Shiny red or orange berries ⅛ inch (4 mm) wide. RANGE: Southeastern U.S.; tropical America. NOTE: All parts are poisonous.

157. Heller's Plantain, Cedar Plantain,
Plantago helleri
Plantaginaceae, Plantain Family

Low hairy annual often less than 6 inches (1.5 dm) tall; in sand, gravel, or limestone, in dry washes, on slopes and plateaus. FLOWERS: Spring; white, with reddish brown center; 4 translucent papery petals; petal lobes to ⅛ inch (4 mm) long; flowers in dense spike to 1½ inches (4 cm) long, held either within or several inches above the leaves; sepals hairy. LEAVES: Simple, basal; blades slender, linear to oblanceolate, 2–8 inches (5–20 cm) long; covered with soft hairs; margins entire. RANGE: Texas; Mexico. RELATED SPECIES: 12 other species in Texas, most with inconspicuous flowers.

158. Anemone, Windflower, *Anemone berlandieri*
(A. heterophylla)
Ranunculaceae, Buttercup Family

Perennial 6–20 inches (1.5–5 dm) tall; in sandy or limestone soil, in lawns, fields, openings in woods; common. FLOWERS: Late winter–early spring; white, pink, blue, or purple; 1–1¾ inches (2.5–4.5 cm) in diameter, solitary at top of hairy stalk; 10–20 linear tepals. LEAVES: Basal, compound, with 3 leaflets; leaflets toothed and often palmately 3-lobed; 3 deeply lobed leaflike bracts in a whorl below and close to flower. FRUIT: Tiny achenes attached to silky hairs; in dense cylindrical spike ¾–1½ inches (2–4 cm) long or more; stalk below fruit often much longer than when in flower. RANGE: Throughout much of Texas; south-central US. RELATED SPECIES: *Anemone caroliniana*, in the eastern half of Texas, has a fruiting head about ½ inch (12 mm) tall. *A. edwardsiana*, endemic to the Edwards Plateau and one variety possibly endangered, usually has 2 or more flowers per stalk.

159. Bedstraw, Cleavers, Stickygrass, *Galium aparine*
  Rubiaceae, Madder Family

Trailing annual weed; rough foliage clings to clothes; in lawns and fields, on gentle slopes in woods, and along seashores; abundant. FLOWERS: Spring; white; minute and inconspicuous. LEAVES: Whorls of 6–8 linear to oblanceolate leaves on succulent square stem; leaves often less than 1 inch (2.5 cm) long but can be 3 inches (7.5 cm); covered with prickly hairs. FRUIT: Tiny 1-seeded ball coated with prickly hairs. RANGE: Native of Eurasia, now widespread in North America. RELATED SPECIES: 18 native species in Texas, some rare or endemic; most occur in restricted areas of East Texas or mountains of the Trans-Pecos. NOTE: The tender new leaves and stems can be cooked and eaten. The fruits can be roasted and ground for a coffee substitute.

160. Lizard's Tail, *Saururus cernuus*
  Saururaceae, Lizard's Tail Family

Perennial to 3 feet (9 dm) tall; in water or wet soil, by streams and lakes, in ditches and marshes; in large colonies. FLOWERS: Spring–fall; fragrant; white; minute, lacking petals; in dense spikes 4–12 inches (1–3 dm) long, tip drooping in flower, becoming erect in fruit. LEAVES: Simple, alternate; blade heart-shaped, 3–6 inches (7.5–15 cm) long; margins entire. RANGE: Eastern U.S.; Canada.

161. Foxglove, *Penstemon cobaea*
  Scrophulariaceae, Figwort Family

Erect perennial to 2½ feet (7.5 dm) tall; on rocky calcareous soil, in prairies and fields, along creeks, on hillsides; common.

FLOWERS: Spring; white, pink, or pale purple; petals united into inflated blossom 1½–2¼ inches (3.5–6 cm) long with bilateral 5-lobed opening; purple lines on interior; 5 stamens, 1 sterile and coated with yellow hairs. LEAVES: Simple, opposite; blade elliptic, lanceolate, or oblanceolate, to 3½ inches (9 cm) long at midstem; surface shiny, hairy (hand lens); margins toothed. RANGE: Nebraska to Texas. RELATED SPECIES: Above characters distinguish this from 22 other species. See scarlet penstemon, page 53, and compare with false dragon's head, page 42.

162. Jimsonweed, Toloache, *Datura wrightii*
Solanaceae, Nightshade Family

Bushy annual 2–4 feet (6–13 dm) tall, from woody root; in dry sand, moist lowlands; uncommon. FLOWERS: Spring–fall; white, may be tinged with purple; large trumpet-shaped flower 4–9 inches (1–2.3 dm) long. LEAVES: Simple, alternate; egg-shaped, 4–10 inches (1–2.5 dm) long; disagreeable odor, bottom usually velvety; margins entire or with shallow, wavy lobes. FRUIT: Nearly round capsule about 1½ inches (4 cm) wide, covered with spines; pendant. RANGE: Texas to California; Mexico. RELATED SPECIES: *Datura inoxia*, in South, Central, and West Texas, is very similar, but mature leaves are not velvety on the bottom; *D. quercifolia*, in West Texas, and *D. stramonium*, an ornamental species, have smaller flowers and erect capsules. NOTE: All parts of *Datura* species deadly poisonous. ❧

7

Vines

163. Poison Ivy, Poison Oak, Hiedra,
Toxicodendron radicans
(Rhus toxicodendron)
Anacardiaceae, Sumac Family

Low shrub or woody, ropy vine, often with aerial roots; in full or partial shade, in dry or moist soil, trailing or shrubby beneath trees and bushes, or climbing high in trees and along fences; abundant and weedy, with several varieties. LEAVES: Deciduous, alternate, pinnately compound, with 3 (rarely 5) leaflets; young leaves often reddish, shiny; fall leaves turn sunset colors; margins extremely variable—entire, coarsely toothed, or shallowly or deeply lobed. FLOWERS: Spring; white or creamy; tiny, inconspicuous. FRUIT: white or creamy waxy berry, to ¼ inch (7 mm) wide. RANGE: Throughout most of Texas; most of Canada and U.S., south to Guatemala; introduced weed in Asia. RELATED SPECIES: *T. radicans* has several sub-species and overlaps extensively in range with 2 similar species, *T. rydbergii* (in the Trans-Pecos) and *T. pubescens*. Poison sumac (*T. vernix*), a small tree in bogs and wetlands of East Texas (mainly between Shelby and Jasper counties), has whitish berries, pinnately compound leaves with 5–13 leaflets, and entire margins. The nonpoisonous sumacs have red fruit (see *Rhus*, p. 172). NOTE: Poison Ivy is sometimes confused with the seedlings of box elder (p. 171) and hop tree (p. 270). Allergic dermatitis is caused by contact with oil from all parts of the poison ivy and poison sumac. Burning the plants allows oil to become airborne. If you suspect contact, wash clothing and skin immediately with soap and water or with aloe vera juice or other astringent (ask pharmacist). Treat a mild rash with the same or with a solution of 1 part bleach to 9 parts water. Contact a doctor if a severe rash develops.

164. Milkweed Vine, *Matelea reticulata*
Asclepiadaceae, Milkweed Family

   Herbaceous perennial vine with milky sap, hairy stems and leaves; in partial shade in woods and thickets, in dry rocky or sandy soil; climbing on fences, shrubs, and small trees; uncommon. LEAVES: Simple, opposite; blade heart-shaped, 1–4½ inches (2.5–11.5 cm) long; margins entire. FLOWERS: Spring–fall; green; flat, star-shaped, ½–¾ inch (1.2–2.0 cm) wide, with silver knob in center; inconspicuous but beautiful; intricate system of veins covers the 5 egg-shaped petals. FRUIT: pod, splitting open to expose flattened seeds attached to long, silky hairs. RANGE: Texas; Mexico. RELATED SPECIES: Above characters distinguish this from 12 other Texas species. Of six endemic species, 2 may be endangered.

165. Cross Vine, *Bignonia capreolata*
Bignoniaceae, Catalpa Family

   High-climbing woody vine with tendrils; on trees in moist woods; uncommon. LEAVES: Evergreen, opposite, compound, with 2 leaflets; blades elliptic or lanceolate, 1–6 inches (2.5–15 cm) long; margins entire. FLOWERS: Spring; red-orange, lighter or yellow inside; to 2 inches (5 cm) long; trumpet-shaped with 5 lobes; in clusters of 2–5 blossoms. FRUIT: Linear leathery pod 4–7 inches (1–1.8 dm) long; seeds winged. RANGE: Eastern U.S., north to New Jersey. NOTE: Widely used for landscaping.

166. Trumpet Creeper, *Campsis radicans*
Bignoniaceae, Catalpa Family

Woody high-climbing vine, clinging to trees and walls with aerial rootlets (tendrils absent); in woods and thickets, along fences, in vacant lots and alleyways; common. LEAVES: Deciduous, opposite, pinnately compound; with about 5–11 leaflets; leaflets 1–3½ inches (2.5–9 cm) long; margins coarsely toothed. FLOWERS: Spring–fall; orange, red-orange, or red; 2–3½ inches (5–9 cm) long; trumpet-shaped, with 5 lobes; in clusters. FRUIT: Leathery or woody pod 2–12 inches (5–30 cm) long; seeds winged. RANGE: Southeastern U.S., north to New Jersey. NOTE: The plant can cause contact dermatitis in some individuals. RELATED SPECIES: *Campsis grandiflora*, in cultivation, lacks aerial roots.

167. Japanese Honeysuckle, *Lonicera japonica*
Caprifoliaceae, Honeysuckle Family

Low to high-climbing woody to shrubby vine; in sun or shade; covering the ground, fences, shrubs, trees; abundant and weedy. LEAVES: Evergreen, simple, opposite; blades oblong, elliptic, or lanceolate, 2–3½ inches (5–9 cm) long; margins entire or occasionally with a few teeth or lobes. FLOWERS: Spring–summer; fragrant; white (may be purple-tinged) turning yellow with age; tubular, with wide 2-lipped mouth; about 1½ inches (4 cm) long. FRUIT: round black berries ¼ inch (7 mm) wide. RANGE: Native of Asia, widely used in landscaping; escaping from cultivation, often displacing native vegetation in the eastern U.S.

168. Coral Honeysuckle, *Lonicera sempervirens*
    Caprifoliaceae, Honeysuckle Family

Shrubby, climbing, or trailing woody vine; in woods, thickets; uncommon. LEAVES: Nearly evergreen, simple, opposite, pair below flowers often united at base; blades oblong, elliptic, or egg-shaped, to 2½ inches (6.5 cm) long; leathery; surfaces usually hair-less or hairs inconspicuous; bottom whitish; margins entire. FLOW-ERS: Nearly year-round, mainly in spring; scarlet (sometimes orange or yellow); slender, trumpet-shaped, 1–2¼ inches (2.5–6 cm) long; in whorls. FRUIT: Red or orange berries to ⅜ inch (1 cm) long; not edible. RANGE: Eastern U.S. NOTE: Beautiful in landscaping. RELATED SPECIES: The rare *Lonicera arizonica*, in the Guadalupe Mountains, has red flowers and hairy leaves; *Lonicera albiflora* has white flowers (page 181).

169. Dodder, *Cuscuta* **species**
    Convolvulaceae, Morning Glory Family

Leafless annual vines; delicate, yellow or orange; parasitic and twining on grasses, wildflowers, trees, some nonflowering plants, and some agricultural crops; some species abundant, completely covering and killing host plant. FLOWERS: Summer–fall; white; tiny. FRUIT: Minute capsules. RANGE: Throughout Texas and U.S. NOTE: 18 species in Texas, some showing a host preference. Lacking green chlorophyll for food production, the parasite must extract its nutrients from another plant. The vines can be used to dye wool yellow.

170. Wild Morning Glory, *Ipomoea trichocarpa*
    Convolvulaceae, Morning Glory Family

Herbaceous low-climbing perennial vine; in sunny areas; fields and gardens, along roadsides and edges of thickets; common and weedy, with 2 varieties. LEAVES: Simple, alternate; blade 1–3 inches (2.5–8 cm) long, shape variable, heart- or egg-shaped; margins with 3–5 lobes or entire. FLOWERS: Spring–fall; rose to purple, with dark center (rarely white); broad trumpet to 2 inches (5 cm) long. RANGE: Southeastern U.S.; Mexico. RELATED SPECIES: 37 other species in Texas.

171. Wood Rose, Alamo Vine,
   Correhuela de las Doce,
   Merremia dissecta (Ipomoea sinuata)
   Convolvulaceae, Morning Glory Family

Herbaceous perennial trailing or climbing vine; in disturbed ground, fields, and open woods, along streams; uncommon. LEAVES: Simple, alternate; blade deeply palmately lobed, 1½–6 inches (4–15 cm) wide; margins of lobes with wavy teeth or lobes. FLOWERS: Spring–fall; white, with purple-red center; open trumpet to 2 inches (5 cm) wide. FRUIT: Round or egg-shaped capsule; when capsule dries, the 5 spreading sepals and capsule become papery and brown, resembling a star or a wooden rose about 2 inches (5 cm) wide; seeds black. RANGE: Texas, Florida; West Indies, Mexico, South America.

172. Buffalo Gourd, Stinking Gourd,
   Calabacilla Amarga, *Cucurbita foetidissima*
   Cucurbitaceae, Gourd Family

Perennial fast-growing vine with long, robust stems sprawling on ground; foliage stinks when crushed; in dry sunny areas, in disturbed soil, along roadsides, in fields; abundant. LEAVES: Simple, alternate; blade to 1 foot (3 dm) long, shape variable, generally triangular or heart-shaped; surface very rough; margins toothed, sometimes shallowly lobed. FLOWERS: Late spring–summer; yellow; large; trumpet-shaped. FRUIT: Summer; round gourd 2–3 inches (5–7.5 cm) wide; dark green with light stripes and blotches, maturing yellow or tan; bitter and inedible. RANGE: California nearly to Mississippi River, north to Nebraska; Mexico. NOTE: Starch in roots and oil in seeds make the plant a potentially valuable crop for the future. RELATED SPECIES: *Cucurbita pepo* var. *texana*, rare, has thin, barely rough leaves and egg-shaped or oval gourds. *C. digitata*, in the Trans-Pesos.

173. Balsam Gourd, *Ibervillea lindheimeri*
    and *I. tenuisecta*
    Cucurbitaceae, Gourd Family

Herbaceous perennial vine with bright red-orange fruits; climbing on shrubbery and fences, at edges of thickets and in open woods; uncommon. LEAVES: Simple, alternate; blade highly variable in shape; margins usually with 3 or 5 shallow or deep lobes, lobes narrowly linear (to ³⁄₁₆ inch (5 mm) wide, in *I. tenuisecta*) or broad (wider than ⅜ inch [1 cm] in *I. lindheimeri*); margins toothed or wavy. FLOWERS: Spring–fall; yellow; tiny, ¼–½ inch (6–12 mm) wide; short tube with 5 narrow lobes; dioecious. FRUIT: Smooth, bright red-orange globe ½–2 inches (1.5–5 cm) in diameter; not edible. RANGE: Texas, Oklahoma. RELATED SPECIES: *Ibervillea tripartita*, endemic to South Texas, has very narrow lobes (may be a variety of *I. lindheimeri*)

174. Chinese Wisteria, *Wisteria sinensis*
    Fabaceae, Legume Family

Thick climbing woody vine; popular in landscaping. LEAVES: Deciduous, alternate, pinnately compound; leaflets oblong or elliptic, 1–3 inches (2.5–8 cm) long; margins entire. FLOWERS: Spring; very fragrant; violet (a white form available); pea-type flowers, numerous in large, showy, drooping racemes. FRUIT: Flattened bean pod 3–6 inches (8–15 cm) long, covered with velvety hairs. RANGE: Native of China, occasionally escaping cultivation. NOTE:

The flowers, pods, seeds, and bark are poisonous. Vines used in basket making. RELATED SPECIES: *Wisteria frutescens*, native to East Texas forests, has smooth pods.

175. Krameria, Trailing Ratany,
Krameria lanceolata
Krameriaceae, Ratany Family

Ground-hugging or trailing perennial; covered with silky hairs; in sandy or dry, rocky soil, in open woods, prairies, fields, hills; uncommon, inconspicuous. FLOWERS: Spring–fall; reddish-purple, ½–¾ inch (1.2–2 cm) wide; 4 or 5 colorful, petallike sepals, much larger than the inconspicuous, unequally shaped petals. LEAVES: Simple, alternate (opposite just below flower); linear to narrowly lanceolate, to 1 inch (2.5 cm) long; margins entire. FRUIT: Round hairy pod ⅜ inch (1 cm) wide, covered with spines. RANGE: Kansas to Arizona and Texas; Mexico. RELATED SPECIES: 3 shrubby species. See range ratany, *Krameria erecta*, page 251.

176. Greenbriar, Catbriar, Zarzaparrilla,
Smilax bona-nox
Liliaceae, Lily Family

Woody climbing or trailing green vine lined with needle-sharp spines, more numerous near base; in thickets and open woods, along fences, in fields; abundant and weedy. LEAVES: Deciduous or persistent, simple, alternate; blade 2–6 inches (5–15 cm) long or longer; shape generally triangular or heart- or egg-shaped; older leaves leathery, sometimes blotched with white; margins entire or lobed, may be spiny. FLOWERS: Spring; greenish; tiny; in round clusters. FRUIT: Fall; round blue-black berry about ¼ inch (7 cm) wide; not edible. RANGE: Southeastern U.S.; Mexico. NOTE: Berries and roots yield dyes for wool, new leaves are edible. RELATED SPECIES: 9 other species, mostly distinctive from this.

177. Carolina Jessamine, *Gelsemium sempervirens*
Loganiaceae, Logania Family

High-climbing perennial vine with slender smooth stems; in moist sandy soil, in trees, at edges of woods; uncommon. LEAVES: Nearly evergreen; simple, opposite; blades elliptic, egg-shaped, or lanceolate, 1–3 inches (2.5–7.5 cm) long; margins entire. FLOWERS: Late winter–spring; fragrant, yellow; 1–1½ inches (2.5–4 cm) long; trumpet-shaped with 5 lobes. FRUIT: Elliptic pod about ¾ inch (2 cm) long; seeds winged. RANGE: Southeastern U.S., north to Virginia; Mexico. NOTE: popular in landscaping; flowers and foliage deadly poisonous.

178. Snailseed, *Cocculus carolinus*
Menispermaceae, Moonseed Family

Herbaceous perennial vine; climbing on shrubs, trees and fences, in woods and thickets, along roadsides and alleys; common. LEAVES: Simple, alternate; blade heart-shaped or triangular, 2–4 inches (5–10 cm) wide and long (or larger), underside softly hairy; margins entire or with wavy lobes. FLOWERS: Summer; greenish white; tiny. FRUIT: Fall; shiny red succulent drupes, about ¼ inch (8 mm) in diameter; seeds coiled like a snail. RANGE: Southeastern U.S., north to Virginia. NOTE: Used in landscaping. The fruit is poisonous. RELATED SPECIES: *Cocculus diversifolius*, in South Texas, has purple fruit and leaves that are usually oblong and lack hairs on underside.

179. Yellow Passionflower, *Passiflora affinis*
Passifloraceae, Passionflower Family

Herbaceous perennial vine; on vegetation and boulders, in rocky areas, along streams; uncommon. LEAVES: Simple, alternate; blade 1–5½ inches wide (2.5–14 cm), palmately 3-lobed, lobes extend at least ⅓ the distance to base of leaf. FLOWERS: Spring–summer; yellow-green; to 1 inch (2.5 cm) wide; as elaborate in design as *Passiflora incarnata*; crown parts threadlike, tipped with knobs; a few minute (⅟₁₆ inch, or 2 mm), deciduous stipules along flower stalk. FRUIT: Blue-black capsulelike berry to ⅜ inch (1 cm) wide; not useful for juice. RANGE: Texas; Mexico. RELATED SPECIES: *Passiflora lutea*, in East and Central Texas, has crown parts that are linear (less threadlike) and not knobby-tipped, and the flower stalk lacks stipules; 3 other similar species.

180. Passionflower, Pasionaria,
  Passiflora incarnata
  Passifloraceae, Passionflower Family

Herbaceous perennial climbing vine; on fences and vegetation, in fields and openings in woods, along roadsides and streams; uncommon. LEAVES: simple, alternate; blade 2–6 inches (5–15 cm) long, deeply palmately 3-lobed; margins toothed. FLOWERS: Spring–summer; purple, pink, and white; elaborate blossom to 3 inches (7.5 cm) wide, with colorful sepals, petals, and threadlike strands emerging from crown. FRUIT: Yellow-orange berry 1–2 inches (2.5–5 cm) long. RANGE: Southeastern U.S.; Bermuda. NOTE: Pulp around seeds is sweet and edible. The plant is cultivated as an ornamental and a source of juice. RELATED SPECIES: *Passiflora foetida*, in South Texas, has smaller yellow to red fruit enclosed by greatly dissected, lacerated sepals, and the leaves are velvety and bad-smelling.

181. Old Man's Beard, Barbas de Chivato,
  Clematis drummondii
  Ranunculaceae, Buttercup Family

Herbaceous or slightly woody perennial vine, covering fences, shrubs, and rocks; in sunny dry areas, and in rocky canyons;

abundant. LEAVES: Opposite, pinnately compound (may be simple near vine tips); leaflets ½–1½ inches (1.2–4 cm) long; margins toothed and usually with 3 or more lobes. FLOWERS: Spring–summer; white or yellowish green, tiny; narrow sepals and sterile stamens resembling petals, about ½ (1.2 cm) long; petals absent; dioecious. FRUIT: Seeds attached to silky hairs, in showy tasslelike clusters. RANGE: Texas to Arizona; Mexico. NOTE: The foliage can irritate the skin. RELATED SPECIES: Above characters distinguish this from 10 other Texas species.

182. Purple Leatherflower, *Clematis pitcheri*
Ranunculaceae, Buttercup Family

Herbaceous perennial vine, sprawling or climbing on shrubs; in shade, in moist soil, in thickets, in open woods, and along waterways; uncommon. LEAVES: Opposite, compound, with 3 or 4 pairs of leaflets; leaflets 1–4 inches (2.5–10 cm) long, shape variable, with deep lobes or again divided into 3 segments; net venation conspicuous. FLOWERS: Spring–summer; purple to brick red outside, darker red or greenish inside; bell-shaped with tips spreading or curling; about 1 inch (2.5 cm) long; 4 sepals, showy, thick, leathery and ribbed; petals absent. FRUIT: Clusters of seeds attached to long, silky hairs. RANGE: Central U.S., north to Indiana. RELATED SPECIES: 10 other species, some similar.

183. Scarlet Leatherflower, *Clematis texensis*
Ranunculaceae, Buttercup Family

Herbaceous or slightly woody perennial vine, low-climbing; in shade, in limestone soil, usually along waterways; uncommon or rare. LEAVES: Opposite, compound, with 3–5 pairs of leaflets; leaflets variable, elliptic, egg-shaped, or nearly round; glaucous; margins entire or with 2 or 3 lobes. FLOWERS: Spring; scarlet; bell-shaped, with tips slightly curled; about 1 inch (2.5 cm) long; 4

sepals, showy, thick, and leathery; petals absent. FRUIT: Clusters of seeds attached to silky hairs, forming a showy feathery ball. RANGE: Endemic to Central Texas. RELATED SPECIES: Above characters distinguish this from 10 other species.

184. Rattan Vine, Supplejack, *Berchemia scandens*
Rhamnaceae, Buckthorn Family

Woody, high-climbing vine; bark smooth, green or reddish brown on new growth; large vines may wrap around trees; in woods; common. LEAVES: Deciduous, simple, alternate; blades oblong, elliptic, or egg-shaped, 1–4 inches (2.5–10 cm) long; with parallel veins conspicuous on bottom; margins entire or slightly scalloped. FLOWERS: Spring; greenish; inconspicuous. FRUIT: Oblong juicy drupe ¼ inch (6 mm) long, blue-black when mature; poisonous. RANGE: southeastern U.S.; southern Mexico, Guatemala. NOTE: The vines have been used in wicker objects. Leaves and stems make good dyes.

185. Southern Dewberry, Zarzamora,
Rubus riograndis (R. trivialis)
Rosaceae, Rose Family

Trailing (sometimes low-arching) perennial vine; stems lined with hard prickles and often with gland-tipped bristly hairs; in thickets, along fences, railroad tracks, and stream banks; abundant and weedy. LEAVES: Persistent, alternate, palmately compound, with 3 or 5 leaflets; leaflet 1–4 inches (2.5–10 cm) long; midrib lined with prickles; margins coarsely toothed. FLOWERS: Late winter–spring; white; 5 delicate wrinkled petals. FRUIT: Spring–summer; blue-black aggregate fruit to 1¼ inches (3 cm) long; juicy, sweet, edible raw or cooked. RANGE: Southeastern U.S. RELATED SPECIES: 9 other species of dewberries and blackberries in Texas.

186. Balloon Vine, Farolitos,
Cardiospermum halicacabum
Sapindaceae, Soapberry Family

Slender herbaceous vine, sprawling along ground and over brush; in dry disturbed soil, in fields, edges of woods; uncommon. LEAVES: Alternate, compound, usually with 3 leaflets divided again into 3 parts; leaflets egg-shaped or lanceolate, ½–3 inches (1.3–8 cm) long, margins coarsely toothed or lobed. FLOWERS: Summer–fall; white; tiny. FRUIT: 3-chambered inflated papery pod 1–1¾ inches (2.5–4.5 cm) wide; seed round, hard, ³⁄₁₆–⅜ inch (5–10 mm) wide, black with white curving scar. RANGE: Widespread in warm areas of Western Hemisphere. RELATED SPECIES: *Cardiospermum corindum*, in West and South Texas, has densely hairy foliage, and a nearly round seed scar; *C. dissectum*, in far South Texas, has leaflets dissected into narrow segments less than ½ inch (1.2 cm) long and ⅛ inch (4 mm) wide.

187. Snapdragon Vine, *Maurandella antirrhiniflora*
(Maurandya antirrhiniflora)
Scrophulariaceae, Figwort Family

Herbaceous perennial vine; sprawling or climbing over fences, shrubs, rocks, beaches, dry salt marshes, dry rocky hills; uncommon. LEAVES: Simple, alternate or opposite; blade triangular or lobed, somewhat like an arrow; ½–1 inch (1.2–2.5 cm) long. FLOWERS: Late winter–fall; violet or purple (rarely completely white), with creamy yellow center; bilateral, resembling miniature snapdragons, to 1 inch (2.5 cm) long; lower lip with hairy interior. FRUIT: Round capsule about ¼ inch (6 mm) wide. RANGE: Texas to California; Mexico. RELATED SPECIES: *Epixiphium wislizeni*, in South Texas and the Trans-Pecos, has light blue flowers, leaves to 2 inches (5 cm) long, and a capsule to ½ inch (1.2 cm) wide.

188. Peppervine, *Ampelopsis arborea*
Vitaceae, Grape Family

Slender woody climbing vine with or without tendrils; in partial shade, along streams and fencerows, in woods and thickets; common. LEAVES: Deciduous, alternate, twice pinnately compound; leaflets ½–2½ inches (1.2–6.5 cm) long; margins with large sharp-pointed teeth, sometimes lobed. FLOWERS: Summer; greenish; tiny. FRUIT: Succulent berries ¼–½ inch (6–15 mm) wide, reddish to black; in clusters; not edible. RANGE: Southeastern U.S.; Mexico. NOTE: Causes contact dermatitis in some individuals. Leaves and tips make excellent dyes.

189. Heartleaf Ampelopsis, *Ampelopsis cordata*
Vitaceae, Grape Family

Slender woody climbing vine with or without tendrils; in woods, usually along waterways; uncommon. LEAVES: Deciduous, simple, alternate; blade heart-shaped or broadly egg-shaped, 2–6 inches (5–15 cm) long and wide; margins with coarse teeth, rarely shallowly 3-lobed. FLOWERS: Spring; greenish; tiny. FRUIT: Berries to ⅜ inch (1 cm) wide, reddish to blue-black; in clusters; not edible, have horrid taste. RANGE: Eastern U.S., north to Ohio; Mexico. NOTE: Causes contact dermatitis in some individuals. RELATED SPECIES: to distinguish from grapes, look for the following: *Ampelopsis* fruit is slightly flattened at the ends and the bark is lined with lenticels and is not shredding.

190. Cow-itch Vine, Hierba del Buey,
    Cissus trifoliata (C. incisa)
    Vitaceae, Grape Family

Herbaceous or partly woody vine, sprawling on ground or climbing with tendrils; in open woods, thickets, marshes; common.

LEAVES: Alternate; blade variable, compound with 3 leaflets or just deeply 3-lobed (occasionally unlobed); to 3 inches (7.5 cm) long; fleshy; fetid when crushed; margins with coarse teeth. FLOWERS: Spring–summer; greenish; tiny. FRUIT: Berries to ⅜ inch (1 cm) wide, black when mature; in loose clusters; poisonous. RANGE: Texas to Florida, north to Kansas, into Mexico. NOTE: Causes contact dermatitis in some individuals.

191. Virginia Creeper, Hiedra,
Parthenocissus quinquefolia
Vitaceae, Grape Family

Woody high-climbing vine; tendrils with 3–8 branches tipped with adhesive disks; in woods, along fences and stream banks; common. LEAVES: Deciduous, alternate, palmately compound with 5 (sometimes 6) leaflets; leaflets 1–6 inches (2.5–15 cm) long, turning red in fall; margins coarsely toothed. FLOWERS: Spring–summer; greenish; tiny; in clusters. FRUIT: Blue-black succulent berries to ¼ inch (7 mm) wide, resembling grapes but deadly poisonous. RANGE: Eastern U.S. NOTE: Causes contact dermatitis in some individuals. RELATED SPECIES: *Parthenocissus vitacea*, in West Texas, has tendrils not as above; *P. heptaphylla*, endemic to the Edwards Plateau, has 7 (sometimes 5 or 6) leaflets that are less than 2 inches (5 cm) long.

192. Mustang Grape, *Vitis mustangensis*
Vitaceae, Grape Family

High climbing vine developing thick trunk and tough woody branches; bark loose, shredding; in various soils, in thickets, along fencerows, stream bottoms, edges of woods; abundant. LEAVES: Deciduous, simple, alternate; blade 2–6 inches (5–15 cm) long and wide, may be heart-shaped or triangular; bottom with dense mat of white feltlike hairs that hide surface; margins toothed, often with

3–5 lobes, sometimes entire. FLOWERS: Spring; greenish; tiny; in clusters. FRUIT: Summer; blue-black grapes ½–¾ inch (1.2–2 cm) wide, in drooping clusters; edible. RANGE: South-central U.S. NOTE: acid of the raw fruit is irritating to the mouth and hands. The fruit is excellent for jellies. RELATED SPECIES: 14 species of grapes in Texas. In other species with a hairy leaf, the bottom surface is visible through the hairs. ❧

8

Trees and Shrubs

QUICK KEY TO TREES AND SHRUBS

After becoming familiar with the longer tree key, below, you may want to use the following quick key.

Leafless trees and shrubs .... go to 116
Trees and shrubs with simple leaves .... go to 25
Leaves simple, opposite, entire .... go to 27
Leaves simple, opposite, toothed or lobed .... go to 41
Leaves simple, alternate .... go to 52
Leaves compound .... go to 6
Leaves compound, with only 3 leaflets .... go to 21

KEY TO TREES AND SHRUBS

NOTE: In most cases this key goes to genus only. Within the text, some genera have keys to species. Refer to poison ivy photograph and description, pages 82 and 133 before using the key.

Poison ivy—see warning
on page 133

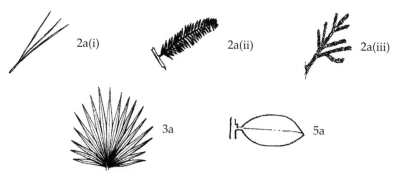

1a. Trees or shrubs without leaves (or apparently so) in all seasons
. . . . go to 116
1b. Trees or shrubs with leaves in most or all months of the year (some
trees lose their leaves in the winter) . . . . go to 2
2a (1b). Trees or shrubs with needles or minute, scalelike leaves . . . .
. . . . one of the following:
   (i) Needles in bundles (fascicles) of 2 or more . . . .
. . . . pine (*Pinus*), p. 260
   (ii) Needles single . . . . one of the following:
      (a) Grows east of the Trans-Pecos . . . .
. . . . bald cypress (*Taxodium*), p. 276
      (b) Found in the Chisos and Guadalupe Mountains . . . .
. . . . blue Douglas fir (*Pseudotsuga menziesii*), p. 260
   (iii) Leaves minute, scalelike . . . . one of the following:
      (a) Berrylike cones to ½ inch (1.2 cm) in diameter, blue or reddish
. . . . juniper (*Juniperus*), p. 218
      (b) Woody cones about 1 inch (2.5 cm) in diameter . . . .
. . . . Arizona cypress (*Cupressus arizonica*), p. 217
      (c) Trees not cone-bearing . . . . tamarisk (*Tamarix*), p. 275
   (iv) Leaves minute, spine-tipped . . . .
. . . . tumbleweed (*Salsola*), p. 215
2b. Not as above . . . . go to 3
3a (2b). Leaves are large palm fronds . . . . palm (*Sabal*), p. 175
3b. Not as above . . . . go to 4
4a (3b). Leaves threadlike or succulent, spine-tipped, minute to 3
       inches (7.5 cm) long; rounded shrub, intricately branching . . . .
. . . . tumbleweed (*Salsola*), p. 215
4b. Not as above . . . . go to 5
5a (4b.) Leaves simple, not compound . . . . go to 25
5b. Leaves compound (when present, leaf bud is at base of whole leaf—use
    position of bud to determine number of leaflets per leaf) . . . go to 6

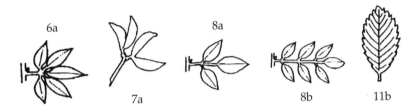

LEAVES COMPOUND

6a (5b). Leaves palmately compound, with 5 or more leaflets . . . .
. . . . buckeye (*Aesculus*), p. 248

6b. Leaves not as above . . . . go to 7

7a (6b). Only 2 leaflets per leaf, leaflets tiny; aromatic desert shrub . . . .
. . . . creosote bush (*Larrea tridentata*), p. 281

7b. More than 2 leaflets per leaf . . . . go to 8

8a (7b). Only 3 leaflets per leaf . . . . go to 21
(See poison ivy, p. 133 before proceeding.)

8b. Three and more leaflets per leaf, leaves pinnately compound . . . .go to 9

LEAVES PINNATELY COMPOUND

9a (8b). Leaflets stiff, margins spine-tipped . . . . *Mahonia*, p. 177

9b. Not as above . . . . go to 10

10a (9b). No thorns at base of leaf or along sides of branches or trunk (though
branches may be rigid and slightly spine-tipped) . . . . go to 13

10b. Thorns along trunk or sides of branches or at base of leaf . . . . . go to 11

LEAVES PINNATELY COMPOUND: THORNS

11a (10b). Fruit a bean pod (legume); leaflets with margins entire (may be
finely toothed on honey locust); flowers in one of the following
forms: (1) strongly bilateral pea-type; (2) slightly bilateral; (3)
minute flowers in dense round heads (puffballs) or cylindrical
spikes (resembling bottle brushes) . . . .
. . . . .legume family (Fabaceae), p. 224–236

11b. Not as above; margins of leaflets toothed; leaflets may have prickles . . . . .
. . . . go to 12

| 13a | 13b | 15a | 15b | 15b(iii) |

12a (11b). Leaves aromatic; fruit a capsule with a black seed . . . .
. . . . prickly ash (*Zanthoxylum*), p. 270

12b. Leaves twice and thrice pinnately compound; fruit a fleshy berry . . . . .
. . . . devil's walking stick (*Aralia spinosa*), p. 175

LEAVES PINNATELY COMPOUND; NO THORNS

13a (10a). Compound leaves opposite (or clusters of compound leaves
opposite) . . . . go to 18

13b. Compound leaves alternate along branches (or clusters of compound
leaves alternate) . . . . go to 14

LEAVES PINNATELY COMPOUND, ALTERNATE; NO THORNS

14a (13b). Fruit a bean pod (legume); leaflets with margins entire (may be
finely toothed on honey locust); flowers in one of the following
forms: (1) strongly bilateral pea-type, (2) slightly bilateral, (3)
minute flowers in dense round heads (puffballs) or cylindrical
spikes (resembling bottle brushes) . . . .
. . . . legume family, (Fabaceae), p. 224–236

14b. Not as above. . . . . go to 15

15a (14b). Margins of leaflets toothed. . . . . go to 16

15b. Margins of leaflets entire (no teeth) . . . . one of the following:

(i) Fruit hard, red, hairy; leaflets small or large, usually aromatic (NOTE: some people are allergic to the oil on fruit
and leaves). . . . . sumac (*Rhus*), p. 172

(ii) Fruit a translucent, yellow, fleshy ball; leaflets 1½–4 inches
(4–10 cm) long . . . . western soapberry (*Sapindus*), p. 272

(iii) Fruit a samara (winged seed); leaflets large, 2–8 inches (5–20
cm) long . . . . tree of heaven (*Ailanthus altissima*), p. 274

17b(ii)   18b   20b(i)   20b(ii)

16a (15a). Fruit a woody or leathery 3-lobed capsule; flowers pink or purple; leaflets large, typically longer than 3 inches (8 cm) . . . .
. . . . Mexican buckeye (*Ungnadia speciosa*), p. 272

16b. Fruit and flowers not as above . . . . go to 17

17a (16b). Fruit a nut with a woody shell, within a husk; leaflets large, longer than 2 inches (5 cm) . . . . one of the following:

> (i) Husk round, enclosing shell, not splitting open; pith of twigs chambered . . . . walnut (*Juglans*), p. 250

> (ii) Husk round, oval, or oblong, splitting open when ripe; pith of twigs not chambered . . . hickory or pecan (*Carya*), p. 249–50

17b. Fruit not as above; leaflets small or large . . . . one of the following:

> (i) Fruit round, yellow, fleshy; flowers lavender to white; leaflets 1–3½ inches (2.5–9 cm) long . . . .
> . . . . chinaberry (*Melia azedarach*), p. 254

> (ii) Fruit a samara (winged seed); leaflets large, 2–8 inches (5–20 cm) long; teeth few or irregularly scattered . . . .
> . . . . tree of heaven (*Ailanthus altissima*), p. 274

> (iii) Fruit hard, red, hairy (NOTE: some people are allergic to oil on fruit and leaves); flowers white; leaflets small or large . . .
> . . . . sumac (*Rhus*), p. 172

LEAVES PINNATELY COMPOUND, OPPOSITE; NO THORNS

18a (13a). Leaflets numerous, tiny, margins entire (no teeth); flowers purple; fruit a small winged capsule, seed encased in scarlet flesh . .
. . . . guayacan (*Guajacum angustifolium*), p. 281

18b. Not as above; leaflets toothed or entire . . . . go to 19

19a (18b). Leaflets large, toothed; branches covered with corky bumps; fruit blue-black, fleshy, in large clusters; flowers white . . . .
. . . . elderberry (*Sambucus*), p. 181

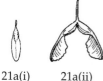

| 21a | 21a(i) | 21a(ii) | 21b | 23a |

19b. Not as above . . . . go to 20

20a (19b). Fruit a cigar-shaped pod filled with winged seeds; flowers showy, yellow to orange; leaflets toothed . . . .
. . . . yellow trumpet flower (*Tecoma stans*), p. 179

20b. Fruit a samara (winged seed); leaflets toothed or entire . . . .
. . . . one of the following:

> (i) Leaflets toothed or entire; single samara, shaped like a canoe paddle . . . . ash (*Fraxinus*), p. 258

> (ii) Leaflets toothed and sometimes lobed; 2 long samaras, joined at base . . . . box elder (*Acer negundo*), p. 171

LEAVES COMPOUND, WITH 3 LEAFLETS

21a (8a). Compound leaves opposite (or clusters of compound leaves opposite); fruit a samara (winged seed) . . . .
. . . . one of the following:

> (i) Single samara, shaped like a canoe paddle; leaflets small, less than 1½ inches (4 cm) long . . . .
> . . . . Gregg ash (*Fraxinus greggii*), p. 258

> (ii) 2 samaras joined at abase; leaflets longer than 2 inches (5 cm) . . . . .box elder (*Acer negundo*), p. 171

21b. Compound leaves alternate along branches (or clusters of compound leaves alternate) . . . . go to 22

22a (21b). Leaflets stiff, margins spine-tipped . . . .
. . . . agarita (*Mahonia trifoliolata*), p. 177

22b. Leaflets not as above . . . . go to 23

23a (22b). Leaflets triangular or 3-lobed, margins not toothed; branches may have spines; flowers and seeds red; fruit a bean pod . . . . . .
. . . . coral bean (*Erythrina herbacea*), p. 228

23b. Not as above . . . . go to 24

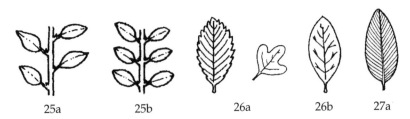

| 25a | 25b | 26a | 26b | 27a |

24a (23b). Leaflets coarsely toothed and may be lobed; aromatic (NOTE: Some people are allergic to oil on fruit and leaves); fruit tiny, red, hairy .... sumac (*Rhus aromatica* or *R. trilobata*), p. 173

24b. Leaflets entire or finely toothed, aromatic; fruit a round, flat samara (winged seed) .... hop tree (*Ptelea trifoliata*), p. 270

LEAVES SIMPLE

25a (5a). Leaves mostly alternate on branches (or clusters of leaves alternate) .... go to 52

25b. Leaves opposite or whorled (or clusters of leaves opposite or whorled) .... go to 26

26a (25b). Leaf margins with teeth or lobes .... go to 41

26b. Leaf margins entire (no teeth or lobes) .... go to 27

LEAVES SIMPLE, OPPOSITE, ENTIRE

27a (26b). Rounded shrub, 3–6 feet (1–2 m) tall, in arid brush in South and Central Texas; leaves less than 3 inches (7.5 cm) long, oblong to elliptic, veins parallel and conspicuous; flowers inconspicuous; fruit fleshy, black ..... coyotillo (*Karwinskia humboldtiana*), p. 264

27b. Not as above .... go to 28

28a (27b). Most leaves longer than 2 inches (5 cm); shrubs to large trees .... go to 36

28b. Leaves generally less than 2 inches long; shrubs or small trees ...go to 29

LEAVES SIMPLE, OPPOSITE, ENTIRE, SHORTER THAN 2 INCHES

29a (28b). Most leaves 1 inch (2.5 cm) long or longer .... go to 33

29b. Leaves generally 1 inch (2.5 cm) or shorter .... go to 30

35b

36a

38a

30a (29b). Shrub with delicate branches; leaves and flowers pleasantly fragrant; flowers white to purple-tinged . . . . bee brush (*Aloysia gratissima*), p. 279

30b. Not as above . . . . go to 31

31a (30b). Small shrub in East Texas; flowers yellow . . . . . *Hypericum*, p. 249

31b. Not as above . . . . go to 32

32a (31b). Shrub in the Trans-Pecos mountains; leaf bottom dotted with minute patches of white hairs (hand lens); leaves narrowly elliptic; flowers white . . . .*Ceanothus greggii*, p. 262
(related species under *C. americanus*)

32b. Small shrub in desert and arid brush; leaves linear to narrowly oblanceolate; flowers tiny, greenish; fruit a purple to black drupe . . . . narrow-leaf forestiera (*Forestiera angustifolia*), p. 257

33a (29a). Flowers inconspicuous, greenish; fruit blue-black; leaves thick and leathery . . . . *Garrya*, p. 247

33b. Flowers showy; not as above . . . . go to 34

34a (33b). Flowers white or pink . . . . one of the following:

(i) Fleshy fruit pink, red, purple, or white *Symphoricarpos*, p. 182

(ii) Berries red or orange . . . . white honeysuckle (*Lonicera albiflora*), p. 181

34b. Flowers red or orange . . . . go to 35

35a (34b). Flowers 5-lobed; fruit red, juicy; stems vining . . . . honeysuckle (*Lonicera*), p. 135

35b. Flowers 4-lobed,with radial symmetry; fruit a round capsule . . . . scarlet bouvardia (*Bouvardia ternifolia*), p. 269

LEAVES SIMPLE, OPPOSITE,
ENTIRE, LONGER THAN 2 INCHES

36a (28a). Leaves linear, less than ½inch (1.2 cm) wide; fruit a long pod . . . . desert willow (*Chilopsis linearis*), p. 179

36b. Leaves broader, not as above .... go to 37

37a (36b). Flowers in white showy ball; shrub near water; leaves in pairs or in whorls of 3 or more .... buttonbush (*Cephalanthus occidentalis*), p. 269

37b. Not as above .... go to 38

38a (37b). Flowers white or pinkish, showy, but not in a tight ball; fruit red, white, or blue; veins of leaf parallel, curving toward tip, conspicuous on bottom .... dogwood (*Cornus*), p. 216

38b. Not as above .... go to 39

39a (38b). Leaves very large, often heart-shaped, in pairs or whorls; fruit a long pod; flowers showy; medium-sized tree .... *Catalpa*, p. 178

39b. Not as above; shrubs or small trees .... go to 40

40a (39b). Leaves thin, 4–8 inches (1–2 dm) long; flowers tassle-like, with 4 linear petals; fruit purple, in loose clusters; in East Texas .... fringe tree (*Chionanthus virginicus*), p. 256

40b. Not as above; leaves somewhat thick and leathery .... one of the following:

    (i) Leaves 2–5 inches (5–12.5 cm) long, entire or finely toothed, bottom usually with rusty hairs; in swamps and other wetlands in East Texas; flowers tiny, numerous; flowers and blue fruit in flat-topped or rounded clusters .... possum haw (*Viburnum nudum*), p. 215

    (ii) Leaves to 3 inches (8 cm) long, bottom often hairy, top smooth or rough; in Central and West Texas; flowers greenish, in inconspicuous catkins; blue-black fruit, few per cluster .... *Garrya*, p. 247

    (iii) Leaves smooth, 2–6 inches (5–15 cm) long; escaping from cultivation in eastern half of state; flowers white, numerous, in large open clusters; blue-black fruit, many per cluster .... *Ligustrum*, p. 259

LEAVES SIMPLE, OPPOSITE, TOOTHED OR LOBED

41a (26a). Most leaves 2 inches (5 cm) long or longer .... go to 46

41b. Most leaves less than 2 inches (5 cm) long .... go to 42

LEAVES SIMPLE, OPPOSITE, TOOTHED, SHORTER THAN 2 INCHES

42a (41b). Flowers inconspicuous, greenish; leaves finely toothed .... spring herald (*Forestiera pubescens*), p. 257

43a   46a   46b   45a, 48a   49a

42b. Not as above; flowers showy . . . . go to 43

43a (42b). Flowers white, red, or lavender, in slender spikes; leaves aromatic
. . . . *Aloysia*, p. 279

43b. Flowers not as above . . . . go to 44

44a (43b). Flowers blue or purple, solitary or few per cluster; leaves aromatic
. . . . shrubby blue sage (*Salvia ballotiflora*) p. 252

44b. Flowers in compact heads or tight rounded clusters, not blue . . . go to 45

45a (44b). Flower heads multicolored; leaves aromatic; fruit blue-black,
poisonous . . . . *Lantana*, p. 280

45b. Flower heads a single color, yellow, orange, or red . . . .
. . . . woolly butterfly bush (*Buddleja marrubiifolia*), p. 253

LEAVES SIMPLE, OPPOSITE, TOOTHED
OR LOBED, LONGER THAN 2 INCHES

46a (41a). Leaves lobed, margins may be toothed or not . . . . go to 49

46b. Leaves toothed but usually not lobed . . . . go to 47

47a (46b). Leaves 3–9 inches (8–23 cm) long; top of leaf rough, bottom velvety; often some leaves deeply lobed . . . .
. . . . paper mulberry (*Broussonetia papyrifera*), p. 254

47b. Not as above . . . . go to 48

48a (47b). Flowers in rounded multicolored heads . . . . *Lantana*, p. 280

48b. Flower clusters 1 color, white to pink . . . . one of the following:

(i) Flower clusters small, inconspicuous; leaves large, 3–9 inches
(7.5–22.5 cm) long, fragrant; fruit in tight clusters, fleshy, pur-
ple or white . . . .
. . . . American beauty berry (*Callicarpa americana*), p. 280

(ii) Flowers and fruit in loose round or flat-topped clusters; fruit
fleshy, blue-black . . . . *Viburnum*, p. 182

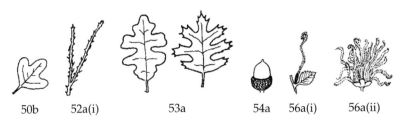

50b   52a(i)   53a   54a   56a(i)   56a(ii)

49a (46a). Margins toothed; leaves 3–9 inches (8–23 cm) long, may be deeply lobed on 1 or both sides or unlobed; top of leaf rough, bottom velvety . . . . paper mulberry (*Broussonetia papyrifera*), p. 254

49b. Not as above . . . . go to 50

50a (49b). Leaves large, often heart-shaped, 6–14 inches (1.5–3.5 dm) long; lobes, if present, are shallow and few, with margins entire; fruit a long pod . . . . *Catalpa*, p. 178

50b. Not as above; leaves palmately lobed . . . . go to 51

51a (50b). Small shrub; flowers white, in showy clusters; leaf margins toothed; fruit blue-black . . . . . . arrowwood (*Viburnum acerifolium*), p. 182

51b. Small to large trees; flowers inconspicuous; margins of lobes toothed or not; fruit a samara with 2 wings . . . . maple (*Acer*), p. 171

LEAVES SIMPLE, ALTERNATE

52a (25a). Bizarre shrub of desert or arid lands; stems emerging from base as vertical wands . . . . one of the following:

　　　(i) Vertical branches 3–30 feet (1–9 m) tall, lined with spines; flowers bright red . . . . . ocotillo (*Fouquieria splendens*), p. 247

　　　(ii) Vertical stems under 3 feet (1 m) tall, fleshy, flexible . . . . leatherstem (*Jatropha dioica*), p. 223

52b. Not as above . . . . go to 53

53a (52b). Leaves deeply pinnately lobed, few to many lobes; no thorns . . . . oak (*Quercus*), p. 238

53b. Leaves not as above; if lobed, lobes shallow or deeply palmately lobed or lobed only on one side of leaf . . . . go to 54

54a (53b). Fruit an acorn . . . . oak (*Quercus*), p. 238

54b. Fruit not an acorn, or fruit not seen . . . . go to 55

55a (54b). Bushy perennial with bright red flowers; leaves 2–5 inches (5–13 cm) long, base heart-shaped, margins toothed, often with 3 shallow lobes in upper half . . . . . Turk's cap (*Malvaviscus*), p. 46

55b. Not as above . . . . go to 56

56a (55b). Seeds attached to feathery plumes ½–3 inches (1.2–8 cm) long; shrubs or small trees in West Texas; flowers white to yellowish; leaves small . . . . one of the following:

    (i) Leaves less than 1 inch (2.5 cm) long, with tiny fingerlike lobes . . . . Apache plume (*Fallugia paradoxa*), p. 266

    (ii) Leaves minute or to 2½ inches (6.5 cm) long, oblong, oval, or obovate; not lobed; margins toothed or not . . . . mountain mahogany (*Cercocapus montanus*), p. 265

    (iii) Leaves to ¼ inch (6 mm) long, linear, spiny, not lobed; margins entire . . . . heath cliffrose (*Purshia ericifolia*), p. 266

56b. Not as above . . . . go to 57

57a (56b). Leaves generally longer than ½ inch (1.2 cm) . . . . go to 59

57b. Leaves minute, generally shorter than ½ inch (12 mm) . . . . go to 58

LEAVES SIMPLE, ALTERNATE, SHORTER THAN ½ INCH

58a (57b). Leaves very narrow, linear to oblong, entire . . . . one of the following:

    (i) Flowers small, reddish purple; small arid land shrub . . . . *Krameria*, p. 251

    (ii) Edges of leaf grooved on bottom; small arid land shrub; branches spine-tipped; flowers inconspicuous; fruit a dry black drupe . . . . javelina bush (*Condalia ericoides*), p. 263

    (iii) Shrub to large tree, near water or in arid land; leaves are minute scales, in some species forming drooping threadlike strands; flowers white to pink, in cylindrical clusters . . . . tamarisk (*Tamarix*), p. 275

58b. Leaves not linear . . . . one of the following:

    (i) Paired spines along branches; leaves toothed or entire, surface usually rough; shrub to small tree in arid brush . . . . granjeno (*Celtis pallida*), p. 277

    (ii) Branches may be spine-tipped; leaf margins entire . . . . similar species under *Condalia ericoides*, p. 263, ... and *C. hookeri*, p. 263

    (iii) No spines; leaves toothed, surface not rough; arid land shrub . . . . snakewood (*Colubrina texensis*), p. 262

 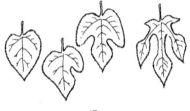 

58a and 59a 65a 66a

LEAVES SIMPLE, ALTERNATE, LONGER THAN ½ INCH

59a (57a). Leaves very narrow (linear to oblong or narrowly elliptic); no spines on branches . . . . go to 62

59b. Leaves broader; branches with or without spines . . . . go to 60

60a (59b). Small shrubs with no spines; white bell-shaped flowers with 5 petals; leaves egg-shaped to nearly round; margins entire or wavy; bottom surface of leaf pale to whitish, densely hairy or smooth . . . . snowbell (*Styrax*), p. 275

60b. Not as above . . . . go to 61

61a (60b). Top surface of leaves rough (may feel like sandpaper) . . . . go to 65

61b. Not as above . . . . go to 71

LEAVES SIMPLE, ALTERNATE, LINEAR

62a (59a). Small evergreen shrub; leaves ⅜–2 inches (1–5 cm) long; margins entire; foliage gray-green or rusty yellow when covered with pollen; seed 4-winged . . . .
. . . . four-wing saltbush (*Atriplex canescens*), p. 215

62b. Not as above . . . . go to 63

63a (62b). Leaves 1–3 inches (2.5–8 cm) long, to ³⁄₁₆ inch (5 mm) wide; margins entire or with few teeth; flowers white; seeds attached to silky hairs; shrub to 10 feet (3 m) tall . . . .
. . . . povertyweed (*Baccharis neglecta*), p. 176

63b. Not as above; leaves 1–6 inches (2.5–15 cm) long or longer; shrub or tree . . . . go to 64

64a (63b). Shrub or small tree; leaves 1–12 inches (2.5–30 cm) long, to ½ inch (1.2 cm) wide, margins entire; showy flowers, white to purple; fruit a long pod, seeds winged; near water or in arid land, not in East Texas . . . . desert willow (*Chilopsis linearis*), p. 179

64b. Not as above .... one of the following:

(i) In moist woods in East Texas; margins entire or with a few teeth; flowers inconspicuous; fruit an acorn ....
.... willow oak (*Quercus phellos*), p. 244

(ii) Near water throughout Texas; margins finely toothed; flowers tiny, in fragrant catkins .... willow (*Salix*), p. 271

LEAVES SIMPLE, ALTERNATE, ROUGH, NOT LINEAR

65a (61a). Leaf margins may be unlobed or deeply palmately lobed or lobed on 1 or both sides; leaves broad, egg-shaped to heart-shaped; margins toothed; leaf bottom smooth, rough, or hairy but not velvety; sap milky; leaf small or large; fruit juicy, red to black or white, resembling a blackberry .... mulberry (*Morus*), p. 255

65b. Not as above .... go to 66

66a (65b). Leaf veins parallel, fairly straight; margins sharply and regularly toothed; fruit an oval flat samara (winged seed); base of leaf uneven on some species .... elm (*Ulmus*), p. 278

66b. Not as above .... go to 67

67a (66b). Branches lined with paired spines; leaves minute to 2 inches (5 cm) long; fruit fleshy, yellow to red .... granjeno (*Celtis pallida*), p. 277

67b. Not as above .... go to 68

68a (67b). Bark sometimes smooth and shiny, etched with short horizontal lines; leaf margins regularly toothed; flowers white, fragrant; branches of some species spine-tipped; fruit fleshy, red purple, or yellow, ⅜ to 1¼ inch (1–3 cm) in diameter ... plum (*Prunus*), p. 267

68b. Not as above .... go to 69

69a (68b). Leaf bottom densely hairy, may be velvety .... go to 70

69b. Not as above; leaves very rough, sandpapery; margins toothed or entire; fruit a yellow to red drupe less than ½ inch (1.2 cm) in diameter; no thorns .... one of the following:

(i) Gray bark may be smooth but typically is covered with raised corky ridges; flowers inconspicuous; leaves deciduous, lanceolate (narrow or so broad as to be nearly heart-shaped); margins toothed or entire; base of leaf often uneven ....
.... hackberry (*Celtis*), p. 277

(ii) Bark furrowed; flowers white, showy; leaves dark green, commonly evergreen; blade broad, elliptic to egg-shaped; margins toothed, wavy, or entire ....
.... anacua (*Ehretia anacua*), p. 180

70a(ii)

70b

72a

70a (69a). Leaves large, longer than 2 inches (5 cm) . . . . one of the following:

> (i) Leaf margins finely toothed or entire, not lobed; flowers white, showy; leaves evergreen; in South Texas . . . .
> . . . . wild olive (*Cordia boissieri*), p. 180

> (ii) Leaf margins toothed; shape variable, unlobed, deeply palmately lobed, or lobed on 1 or both sides; flowers green, inconspicuous; sap milky; leaves deciduous . . . .
> . . . . paper mulberry (*Broussonetia papyrifera*), p. 254

70b. Leaves small, typically less than 2 inches (5 cm) long; margins scalloped . . . . *Bernardia*, p. 223

LEAVES SIMPLE, ALTERNATE, NOT LINEAR, NOT ROUGH

71a (61b). Leaves large, deeply palmately lobed or lobed on 1 or both sides . . . . go to 76

71b. Not as above; if lobed, lobes are shallow . . . . go to 72

LEAVES SIMPLE, ALTERNATE, NOT DEEPLY LOBED

72a (71b). Leaves 3-lobed in upper half, or somewhat triangular, with widened tip and narrow base; leaves 2–7 inches (5–18 cm) long; trees not thorny . . . . . blackjack oak (*Quercus marilandica*), p. 240
. . . . *or* water oak (*Q. nigra*), p. 242

72b. Not as above . . . . go to 73

73a (72b). On most leaves, margins wavy (not sharply toothed) or with shallow rounded lobes; leaves widest near middle or in upper half . . . . one of the following:

> (i) Base of leaf uneven; leaves 2–6 inches (5–15 cm) long; flowers form yellow tassels . . . . . witch hazel (*Hamamelis*), p. 247

> (ii) Leaves small or large; flowers inconspicuous catkins . . . .
> . . . . oak (*Quercus*), p. 238

73b. Not as above . . . . go to 74

73a(i)   73a(ii)

75a

75b

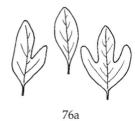

76a   77a   78a

74a (73b). Bark smooth, white to pink or red; older bark peeling in sheets; leaf margins finely toothed or entire; leaves 1–4 inches (2.5–10 cm) long .... madrone (*Arbutus xalapensis*), p. 221

74b. Not as above .... go to 75

75a (74b). Leaf margins toothed or spiny (teeth may be tipped with hairlike bristles) .... go to 98

75b. Leaf margins entire, not toothed or spiny .... go to 79

LEAVES SIMPLE, ALTERNATE, DEEPLY PALMATELY LOBED OR LOBED ON ONE OR BOTH SIDES

76a (71a). Margins of lobes not toothed; tips of lobes not tapering; leaves aromatic; shape variable, unlobed or palmately lobed or lobed on 1 or both sides of leaf .... sassafras (*Sassafras albidum*), p. 252

76b. Not as above .... go to 77

77a (76b). Leaf somewhat star-shaped; margins of lobes toothed; fruit a woody spiny ball .... sweetgum (*Liquidambar styraciflua*) p. 248

77b. Not as above .... go to 78

78a (77b). Leaf margins coarsely toothed; leaf shape variable, unlobed or palmately lobed or lobed on 1 or both sides of leaf ....
.... mulberry (*Morus*), p. 255

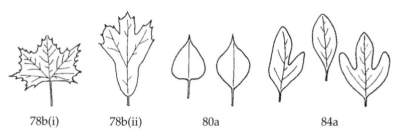

| 78b(i) | 78b(ii) | 80a | 84a |

78b. Not as above . . . . one of the following:

> (i) Leaves typically as broad or broader than long, palmately lobed (resembling a maple leaf); tips of lobes pointed, often tapering; fruit a woody ball; where older bark peels away, new bark smooth, white or creamy . . . .
> . . . . sycamore (*Platanus occidentalis*), p. 261
> (ii) Lobes and teeth tipped with hairlike bristles; typically with 3 lobes in upper half of leaf . . . .
> . . . . southern red oak (*Quercus falcata*), p. 245

LEAVES SIMPLE, ALTERNATE, ENTIRE

79a (75b). Leaves heart-shaped, 2–6 inches (5–15 cm) long; flowers pink to purple; fruit a short flat pod . . . . redbud (*Cercis canadensis*), p. 227

79b. Not as above . . . . go to 80

80a (79b). Leaves triangular (widest near base) or somewhat diamond-shaped, often as wide or wider than long; generally longer than 2 inches (5 cm) . . . . one of the following:

> (i) Tree; flowers in catkins; fruit a 3-lobed capsule with 3 hard white seeds; sap milky . . . .
> . . . . Chinese tallow (*Triadica sebifera*), p. 224
> (ii) Shrub or small tree; young trunk and branches green; leaves lanceolate to triangular; flowers yellow, tubular . . . .
> . . . . tree tobacco (*Nicotiana glauca*), p. 274

80b. Not as above . . . . go to 81

81a (80b). Leaves not strongly aromatic . . . . go to 85

81b. Leaves strongly aromatic . . . . go to 82

82a (81b). In far West Texas; foliage sticky, smelling like creosote . . . .
. . . . tarbush (*Flourensia cernua*), p. 176

82b. Not as above; not in far West Texas . . . . go to 83

83a (82b). Leaves narrow, may widen toward tip, evergreen, smell like bay leaves; fruit hard, about ⅛ inch (4 mm) wide, coated with white wax . . . . wax myrtle (*Morella*), p. 256

83b. Not as above .... go to 84

84a (83b). Leaves thin, deciduous, turning sunset colors in fall, lemony smelling; shape variable, unlobed or deeply palmately lobed or lobed on one or both sides .... sassafras (*Sassafras albidum*), p. 252

84b. Leaves large, thick and leathery, evergreen ..... one of the following:

(i) Leaf bottom covered with rusty hairs; flowers large, white, fragrant; fruit conelike, filled with red seeds .... southern magnolia (*Magnolia grandiflora*), p. 253

(ii) Leaves smell like bay leaves; flowers inconspicuous; fruit blue-black, about ½ inch (1.2 cm) wide .... red bay (*Persea borbonia*), p. 252

85a (81a). Shrub or small tree in sandy soil from East Texas to Bastrop County; leaves to 3 inches (7.5 cm) long, often leathery; flowers tiny, white to pink, urn-shaped; fruit a small dry red to black berry .... farkleberry (*Vaccinium arboreum*), p. 222

85b. Not as above .... go to 86

86a (85b). Many leaves longer than 2 inches (5 cm) .... go to 90

86b. Almost all leaves shorter than 2 inches (5 cm) long .... go to 87

LEAVES SIMPLE, ALTERNATE, ENTIRE, SHORTER THAN 2 INCHES

87a (86b). Branches not spiny .... go to 89

87b. Spines at tips or along sides of branches (in some species, spines not always present on older branches); fruit blue or black .... go to 88

88a (87b). Leaf oblong or widest near base; margins entire or toothed; young branches green or gray-green, often with a whitish waxy coating .... lotebush (*Ziziphus obtusifolia*), p. 265

88b. Not as above .... one of the following:

(i) Leaf widest near tip, base very narrow; spines only at tips of branches .... brasil (*Condalia hookeri*), p. 263

(ii) Leaves elliptic or slightly wider in upper half; sap may be milky; spines at tips or along sides of branches .... Coma (*Sideroxylon*), p. 273

89a (87a). Bark smooth, gray; shrub or small tree; leaf margins often rolled under slightly; fruit black, juicy .... Texas persimmon (*Diospyros texana*), p. 220

| 88b(i) | 88b(ii) | 93a | 98a |

89b. Not as above . . . . one of the following:

   (i) Leaves grayish green, soft, less than 1 inch (2.5 cm) long; flowers showy, purple . . . . cenizo (*Leucophyllum*), p. 273

   (ii) Leaves green or gray, longer than 1 inch (2.5 cm); flowers in inconspicuous catkins . . . . oak (*Quercus*), p. 238

LEAVES SIMPLE, ALTERNATE, ENTIRE, LONGER THAN 2 INCHES

90a (86a). Spines often present on branches; sap sometimes milky . . . .
. . . . one of the following:

   (i) Leaves 1–4 inches (2.5–10 cm) long, bottoms densely hairy in some species; spines at tips of branches or along sides; fruit small, fleshy, purple-black . . . . . . Coma (*Sideroxylon*), p. 273

   (ii) Leaves 1–8 inches (2.5–20 cm) long; stout thorns on sides of branches; wood orange; fruit large, green, resembling a brain . . . . osage orange (*Maclura pomifera*), p. 255

90b. Branches not spiny; not as above . . . . go to 91

91a (90b). Gray bark may be smooth but typically is covered with raised corky ridges; leaves lanceolate (narrow or so broad as to be nearly heart-shaped); margins sparsely toothed or entire; base of leaf sometimes uneven; fruit red or orange . . . . . . hackberry (*Celtis*), p. 277

91b. Not as above . . . . go to 92

92a (91b). Leaf bottom covered with rusty hairs; leaves evergreen, stiff, dark green, 3–10 inches (8–25 cm) long; flowers large, white, fragrant . . . . southern magnolia (*Magnolia grandiflora*), p. 253

92b. Not as above . . . . go to 93

93a (92b). Leaves 2–6 inches (5–15 cm) long, oblong to elliptic; veins parallel, conspicuous on bottom; shrubs or small trees; flowers green, inconspicuous; fruit fleshy, red to black . . . .
. . . . buckthorn (*Rhamnus*), p. 264

93b. Not as above . . . . go to 94

94a (93b). Flowers showy, yellow, white, or pink; shrubs or small trees . . . .
. . . . go to 96

94b. Flowers inconspicuous (if tree is large, flowers may not be visible);
shrubs or small or large trees . . . . go to 95

95a (94b). Leaves 2–12 inches (5–30 cm) long; large trees in or near water in
East Texas; base of trunk enlarged in 1 species; flowers minute;
fruit purple, oblong, less than 1 inch (2.5 cm) long . . . .
. . . . tupelo (*Nyssa*), p. 217

95b. Not as above . . . . one of the following:

(i) Leaves 2–6 inches (5–15 cm) long; medium-sized trees; flowers small, yellowish, bell-shaped; fleshy fruit to 2½ inches
(6.5 cm) wide, yellow, orange, or red . . . .
. . . . common persimmon (*Diospyros virginiana*), p. 220

(ii) Flowers in catkins; shrubs or trees . . . . .oak (*Quercus*), p. 238

96a (94a). In Coastal, South, and Central Texas and into the Trans-Pecos;
flowers yellow, tubular . . . tree tobacco (*Nicotiana glauca*), p. 274

96b. Not as above; found in East Texas . . . . go to 97

97a (96b). Showy flowers trumpet-shaped, pink or white; leaves narrow,
finely toothed or entire . . . . azalea (*Rhododendron*), p. 222

97b. Not as above; flowers white . . . . one of the following:

(i) Leaves to 4½ inches (1.2 dm) long; tiny flowers in short elongated clusters . . . . . laurel cherry (*Prunus caroliniana*), p. 267

(ii) Leaves 4–8 inches (1–2 dm) long; flower petals linear, about
1 inch (2.5 cm) long, tassel-like . . . .
. . . . fringe tree (*Chionanthus virginicus*), p. 256

LEAVES SIMPLE, ALTERNATE, TOOTHED OR SPINY

98a (75a). Leaf tip or teeth spine-tipped . . . . one of the following:

(i) Leaves stiff; fruit round, bright red . . . .
. . . . American holly (*Ilex opaca*), p. 174

(ii) Fruit round, dark red to black . . . .
. . . . laurel cherry (*Prunus caroliniana*), p. 267

(iii) Fruit an acorn . . . . oak (*Quercus*), p. 238

98b. Teeth not spiny (though may be tipped with hairs) . . . . go to 99

99a (98b). Branches not thorny . . . . go to 101

99b. Trees or shrubs with thorns . . . . go to 100

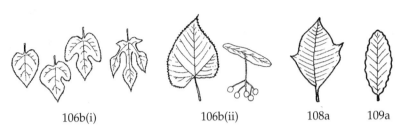

106b(i)   106b(ii)   108a   109a

100a (99b). Thorns at tips of branches . . . . one of the following:

    (i) Thorn-tipped branches short, rigid, green or gray-green, often with whitish waxy coating; fruit to ⅜ inch (1 cm) long . . . . lotebush (*Ziziphus obtusifolia*), p. 265

    (ii) Fruit ⅜–1 inch (1–2.5 cm) long; smooth young bark etched with short, horizontal lines . . . . plum (*Prunus*), p. 267

100b. Stout thorns along sides of branches, not at tips . . . .
. . . . hawthorn (*Crataegus*), p. 266

101a (99a). Gray bark may be smooth but typically is covered with raised corky ridges; base of leaf sometimes uneven; leaves lanceolate (narrow or so broad as to be nearly heart-shaped); margins sparsely toothed or entire; fruit small, round, red or orange . . .
. . . . hackberry (*Celtis*), p. 277

101b. Not as above . . . . go to 102

102a (101b). Leaves narrow, toothed in upper half or entire; leave aromatic, smelling like bay leaves; fruit in clusters of tiny hard balls, coated with whitish wax . . . . wax myrtle (*Morella*), p. 256

102b. Not as above . . . . go to 103

103a (102b). Trees with reddish or gray bark peeling in thin strips; smooth young bark etched with short, horizontal lines; in East Texas; leaves 1–4 inches (2.5–10 cm) long, bottom pale or whitish; fruit conelike . . . . river birch (*Betula nigra*), p. 177

103b. Not as above . . . . go to 104

104a (103b). Small shrub; leaves heart-shaped or triangular with the texture of velveteen, less than 2 inches (5 cm) long; flowers pink, showy . . . . rose pavonia (*Pavonia lasiopetala*), p. 254

104b. Not as above . . . . go to 105

105a (104b). Small to large trees; leaves longer than 2 inches (5 cm), heart-shaped, triangular (widest near base), or ovate, often nearly as wide as long; margins coarsely toothed . . . . go to 106

105b. Not as above; leaves small or large, may be elliptic, oval, or slightly wider at either end .... go to 107

106a (105a). Leaves large, triangular (widest near base) or diamond shaped, 3–7 inches (7.5–18 cm) long; large trees with furrowed bark .... cottonwood (*Populus*), p. 271

106b. Leaves ovate or heart-shaped .... one of the following:

(i) Sap often milky; leaf margins unlobed or deeply palmately lobed or lobed on 1 or both sides; leaf base may or may not be uneven; fruit cylindrical, juicy, resembling a blackberry .... mulberry (*Morus*), p. 255

(ii) Margins not lobed; leaf base usually lopsided; fruit in clusters of tiny balls dangling from leaflike bract .... basswood (*Tilia americana*), p. 276

107a (105b). Small shrub in arid brush; leaves less than 2 inches (5 cm) long, finely toothed; no thorns; branches gray, zigzag; flowers inconspicuous; fruit hard, black .... snakewood (*Colubrina texensis*), p. 262

107b. Not as above .... go to 108

108a (107b). Margins entire or with a few prominent teeth; large trees in or near water in East Texas; base of trunk enlarged in one species; leaves 2–12 inches (5–30 cm) long; fruit purple .... tupelo (*Nyssa*), p. 217

108b. Not as above .... go to 109

109a (108b). Leaf margins with small rounded teeth; leaves less than 3 inches (8 cm) long; fruit bright red; shrub or small tree .... yaupon (*Ilex*), p. 174

109b. Not as above .... go to 110

110a (109b). Bark smooth, gray, not etched with horizontal lines; trees in East Texas .... one of the following:

(i) Leaves 3–6 inches (7.5–15 cm) long; large trunk .... beech (*Fagus grandifolia*), p. 237

(ii) Leaves to 3½ inches (8 cm) long; trunk slender .... ironwood (*Carpinus caroliniana*), p. 178

110b. Not as above .... go to 111

111a (110b). Young bark smooth, but etched with short horizontal lines; bark reddish or gray, shiny on some species; older bark may be rough; on black cherry bark is dark and flaky; fruit fleshy .... plum or cherry (*Prunus*), p. 266–69

| 114a | 115a(ii) | 115b | 117b | 119b |

111b. Not as above . . . . go to 112

112a (111b). Small shrubs with tiny white flowers in small rounded clusters; leaf margins finely toothed; fruit a dry capsule . . . . *Ceanothus*, p. 262

112b. Not as above . . . . go to 113

113a (112b). Shrub in East Texas Forests; spring flowers showy, white to pink; leaves narrow, 1–4 inches (2.5–10 cm) long, entire or finely toothed . . . . azalea (*Rhododendron*), p. 222

113b. Not as above . . . . go to 114

114a (113b). Leaf margins with numerous widely spaced teeth; veins may project from teeth as short bristles or long hairs . . . . one of the following:

(i) Trees, some quite large; leaves 3–10 inches (8–25 cm) long; veins often project from teeth as short bristles; fruit an acorn . . . . chinkapin oak (*Quercus muehlenbergii*, West, Central, and Northwest Texas, in limestone), p. 238, or chestnut oak (*Q. michauxii*, East Texas moist woods), p. 243

(ii) Shrub or tree in sandy woods of East Texas; leaves 3–6 inches (7.5–15 cm) long, teeth usually tipped with hairs; nutlets enclosed in prickly hull . . . . chinquapin (*Castanea*), p. 237

114b. Not as above; teeth small . . . . go to 115

115a (114b). Teeth sharp-pointed; fruit not fleshy . . . . one of the following:

(i) Bark very shaggy; leaves 2–4½ inches (5–12 cm) long; in East Texas woods or Trans-Pecos mountains; tiny nutlets enclosed in tiny papery sacks . . . . hop hornbeam (*Ostrya*), p. 178

(ii) Leaves small or large; various species around Texas; fruit a round flat samara (winged seed) . . . . . elm (*Ulmus*), p. 278

115b. Shrub or small tree; veins parallel and prominent on leaf bottom; fruit fleshy, red to black; leaves to 6 inches (1.5 dm) long, or less than 3 inches (8 cm) on some species; margins finely toothed or entire; flowers green, inconspicuous . . . . buckthorn (*Rhamnus*), p. 264

LEAFLESS TREES AND SHRUBS

116a (1a). No spines or thorns present .... go to 118

116b. Spines or thorns present .... go to 117

117a (116b). Intricately branched thorny tree or shrub; young branches green; leaves usually absent .... allthorn (*Koeberlinia spinosa*), p. 251

117b. Large shrub with vertical wandlike branches, 3–30 feet (1–9 m) tall; branches thorny .... ocotillo (*Fouquieria splendens*), p. 247

118a (116a). Small branching shrubs; young branches green or yellow .... mormon tea (*Ephedra*), p. 221

118b. Unusual arid land plants with stems to 3 feet (1 m) tall, emerging from ground as vertical wands or rods .... go to 119

119a (118b). Stems fleshy, flexible ...... leatherstem (*Jatropha dioica*), p. 223

119b. Stems are stiff rods to 3/16 inch (5 mm) in diameter, coated with wax ... . candelilla (*Euphorbia antisyphilitica*), p. 222

TREES AND SHRUBS

(photos 193 through 195) Maple, *Acer* species (see Key)
Aceraceae, Maple Family

Small to large trees, often dominant forest trees. LEAVES: Deciduous, simple or pinnately compound, opposite; usually palmately 3- 5-lobed. FLOWERS: Late winter-spring; inconspicuous; tiny; wind-pollinated; monoecious or dioecious. FRUIT: 2-winged samara (winged seed). NOTE: Maples may be found as ornamentals outside their natural range; non-native ornamentals are not included here. In northern states, the sap of some species yields syrup. Leaves make nice dyes.

KEY TO MAPLES

1a. Leaves pinnately compound, with 3–9 leaflets ....Box elder, ash-leaved maple, *Acer negundo* (photo 194)—young branches may be green and smooth; common and weedy; often near streams, in East, North Central, and Central Texas; leaflets broad, 2–5 inches (5–13 cm) long; leaf margins coarsely toothed, often palmately lobed; leaflets resemble those of poison ivy.

1b. Leaves simple . . . . go to 2

2a (1b). In scattered populations in canyons and mountains of Edwards Plateau and Trans-Pecos . . . .
Bigtooth maple, *Acer grandidentatum* (several varieties)(photo 193)—Trees small to 60 feet (18 m) tall; leaf blade small, 2–4 inches (5–10 cm) wide, lobes with nipplelike rounded projections or entire; leaves turning sunset colors in fall.

2b. In forests of East Texas . . . . go to 3

3a (2b). On leaf, sides of lobes bear numerous sharp teeth, angle between lobes sharp; branchlets and leaf petioles red; blade 2–6 inches (5–15 cm) long and wide, dark green above, bottom pale and nearly hairless to white and densely hairy; leaves turning red in fall; common; trees small to 70 feet (22 m) . . . .
. . . . red maple, *Acer rubrum* (photo 195) (several varieties)

3b. Sides of leaf lobes smooth or wavy (without sharp teeth), tips of lobes often with nipplelike projections; angle between lobes round . . . . .
. . . . go to 4

4a (3b). Small to large trees 20–105 feet (6–32.3 m) tall; leaves 2–8 inches (5–20 cm) wide, bottom pale, smooth to hairy; common . . . .
. . . . southern sugar maple, *Acer barbatum* (*A. saccharum*)

4b. Small trees less than 20 feet (6 m) tall, leaves usually less than 3½ inches (9 cm) wide, bottom velvety; only in Sabine National Forest and surrounding counties . . . . chalk maple, *Acer leucoderme*

(photos 196 through 199) Sumac, *Rhus* species (see Key)
Anacardiaceae, Sumac Family

Shrubs or small trees, some popular in landscaping. LEAVES: Pinnately compound, alternate; usually aromatic, with lemony smell. FLOWERS: White, greenish, or pale yellow; tiny. FRUIT: Hard red seedlike drupe to ¼ inch (7 mm) long, covered with oily hairs. NOTE: The fruit yields a sour drink when steeped in warm water. Some individuals develop dermatitis from contact with the oil on the plant and fruit. RELATED SPECIES: Poison sumac (*Toxicodendron vernix*) causes severe skin rash; an uncommon shrub or small tree in swamps and wetlands of East Texas; it has tiny white fruits and 5–13 leaflets 2–4 inches (5–10 cm) long, with entire margins. See poison ivy and its related species (page 133). Texas pistache (*Pistacia texana*), a small, rare tree in Val Verde County, has the following characteristics: 9–21 leaflets ½–1 inch (1.2–2.5 cm) long, margins entire; evergreen in mild winters; midrib not winged; fruit red, turning blue-black, hard, waxy, not hairy. Mango and cashew are tropical relatives.

KEY TO SUMACS

1a. Leaves with only 3 leaflets, small shrubs, leaves deciduous . . . . . go to 2

1b. Leaves with 3 and more leaflets . . . . go to 3

2a (1a). Several varieties, throughout state except South Texas; common; leaflet size variable—on some shrubs minute, ¼–½ inch (6–12 mm) long, or 1–3½ inches (2.5–9 cm) long; leaf surfaces hairless; leaf margins coarsely toothed and often lobed . . . .
. . . . Fragrant sumac, *Rhus aromatica* (photo 196)

2b. Leaves densely hairy, velvety, West Texas. . . . .
. . . . Skunkbush sumac, *R. trilobata.*

3a (1b). Leaflets minute, less than ¾ inch (2 cm) long, ¼ inch (6 mm) wide. Littleleaf sumac, correosa, *Rhus microphylla* (photo 198)—Small shrub in dry soil, in western half of Texas, west of Austin; branches stiff, somewhat spine-tipped; leaves deciduous, 5–9 leaflets per leaf, midrib winged, margins entire.

3b. Leaflets larger than above . . . . go to 4

4a(3b). Leaves evergreen, often shiny, leathery; 3–9 leaflets per leaf; leaflets to 2 inches (5 cm) long; margins entire; rounded shrub in dry rocky soil, Central Texas to the Trans-Pecos . . . .
. . . . evergreen sumac, *Rhus virens* (2 subspecies) (photo 199)

4b. Leaves deciduous, turning sunset colors in fall; leaflets generally lance-olate, 1–4 inches (2.5–10 cm) long, 7 to many per leaf; margins entire or toothed; slender shrubs or small trees; flowers and fruit in large showy clusters, erect at branch tips . . . . go to 5

5a (4b). Leaflets mostly sharp-toothed (teeth numerous), lanceolate to oblong, 2–5 inches (5–12.5 cm) long; midrib not winged; 11–31 leaflets per leaf; in East and North Central Texas . . . .
. . . . smooth sumac, *Rhus glabra*

5b. Leaflets mostly with margins entire (sometimes with some teeth); midrib flattened or winged; to 21 leaflets per leaf . . . . go to 6

6a (5b). Leaflets narrowly lanceolate, generally less than ½ inch (1.5 cm) wide, strongly falcate; in Trans-Pecos and Central and North Central Texas, west of Dallas and Austin, and in Palo Duro Canyon . . .
. . . . prairie flameleaf sumac, *Rhus lanceolata* (photo 197)

6b. Leaflets broadly lanceolate to oblong, to 1½ inches (4 cm) wide, not strongly falcate; in East Texas to Austin . . . .
. . . . shining sumac, winged sumac, *Rhus copallinum*

200. American Holly, *Ilex opaca*
Aquifoliaceae, Holly Family

Small to large tree to 58 feet (17.9 m); in moist woodlands; common with 300 horticultural varieties. LEAVES: Evergreen, simple, alternate; blade oval or elliptic to oblong, 2–4 inches (5–10 cm) long; stiff; margins set with sharp pines or entire and spine-tipped. FLOWERS: Spring–summer; white to yellowish; tiny; dioecious. FRUIT: Red drupe (rarely yellow), elliptic or round, to ½ inch (1.2 cm) long; poisonous. RANGE: Eastern U.S. NOTE: This holly is often used in Christmas decorations and is an excellent shade-tolerant landscape tree or pruned hedge.

201. Yaupon Holly, Cassina, *Ilex vomitoria*
Aquifoliaceae, Holly Family

Shrub or small tree to 45 feet (13.9 m); a common understory plant. LEAVES: Evergreen, simple, alternate; blade oval to oblong, ½–2 inches (1.2-5 cm) long; thick, leathery; margins with small rounded teeth. FLOWERS: Spring; whitish; tiny; scattered to densely clustered along branches; dioecious. FRUIT: Red shiny drupes ¼ inch (6 mm) in diameter; poisonous. RANGE: Southeastern U.S. NOTE: Popular in landscaping, yaupon holly has dwarf and yellow-fruited varieties. The caffeine-rich leaves can be dried in the oven and used for tea. RELATED SPECIES: Deciduous yaupon (*Ilex decidua*), or possum haw, has leaves that are slightly larger (to 3 inches, or 8 cm), thinner, and often crowded on the tips of stubby branches; the fruit is red and opaque. Other hollies, some with purple or black fruit, are restricted to narrow ranges in East Texas.

202. Devil's Walking Stick, Hercules' Club,
Aralia spinosa
Araliaceae, Ginseng Family

Shrub or skinny tree, small to 44 feet (13.5 m) tall, with stout thorns on trunk and prickles on leaves; in woods and along streams; common. LEAVES: Deciduous, twice and thrice pinnately compound, alternate; whole leaf to 4½ feet (1.4 m) long; leaflets elliptic to egg-shaped, 1–5 inches (2.5–12.5 cm) long; margins toothed. FLOWERS: Late spring–summer; white or yellowish; tiny; in clusters to 4 feet (1.2 m) long. FRUIT: red to blue-black berry, ¼ inch (7 mm) in diameter; may be poisonous. RANGE: Eastern U.S.

203. Dwarf Palmetto, *Sabal minor*
Arecaceae, Palm Family

Trunkless palm with fronds arising from underground stem; occasionally developing trunk 6–9 feet (2–3 m) tall; scattered populations; usually in swamps, floodplains, moist soil; some hill country populations in drier soil. LEAVES: Evergreen; fan-shaped fronds to 5 feet (1.5 m) or more in diameter; leafstalk without spines. FRUIT: Large clusters of nearly round black fruits, ¼–½ inch (7–12 mm) in diameter. RANGE: Southeastern U.S.

204. Sabal Palm, Texas Palmetto,
Palma de Mícheros, *Sabal mexicana (S. texana)*
Arecaceae, Palm Family

Stately palm to 49 feet (15 m) tall; trunk smooth or covered with woody bases of old fronds; very rare in the wild. LEAVES: Evergreen;

fan-shaped frond, 4–7 feet (1.2–22 m) long, folds upward along down-curved midrib; leafstalk without spines. FRUIT: Large clusters of dull black, thin-fleshed fruit, to ¾ inch (2 cm) in diameter; seed hard, shiny, brown; edible pulp, but fruit of nonnative ornamental palms may be toxic. RANGE: along the Rio Grande and rivers draining into the central coast; upriver limits of pre-settlement range unknown; today found wild in Cameron, Victoria, and Jackson counties; Mexico and Central America. NOTE: Though sabal palms are grown ornamentally, almost all native stands have been cut down. Former range much more extensive. *Sabal minor* and *S. mexicana* are the only native palms in Texas.

205. Povertyweed, Roosevelt Weed,
Jara Dulce, *Baccharis neglecta*
Asteraceae, Sunflower Family

Weedy shrub about 3–10 feet (1–3 m) tall; branches and leaves smooth and essentially hairless; often growing in disturbed areas; widespread and abundant. LEAVES: Deciduous or somewhat persistent, simple, alternate; blade linear to narrowly elliptic, 1–3 inches (2.5–8 cm) long, less than ³⁄₁₆ inch (5 mm) wide; surface may be sticky; margins entire or with few minute teeth. FLOWERS: Late summer–fall; white; in tiny heads to ³⁄₁₆ inch (5mm) long; dioecious. FRUIT: Tiny achene attached to silky hairs. RANGE: Much of southern U.S.; Mexico. NOTE: The common names refer to the plant's rapid expansion onto deserted farmland and ranchland during the Depression. RELATED SPECIES: Leaf width distinguishes this from most of the 10 other species.

206. Tarbush, Hojasé, *Flourensia cernua*
Asteraceae, Sunflower Family

Small shrub about 3–6 feet (1–2 m) tall; resinous sticky foliage smelling like creosote; in desert and dry plains. LEAVES: Persistent in mild winters, simple, alternate; blade elliptic to oval, to 1 inch (2.5

cm) long; margins entire or with few teeth. FLOWERS: Fall; yellow; in tiny heads. RANGE: Texas to Arizona and Mexico. NOTE: One of the most abundant shrubs of the Chihuahuan Desert, tarbush is usually associated with creosote bush (*Larrea tridentata*), see page 281.

207. Agarita, *Mahonia trifoliolata (Berberis trifoliolata)*
Berberidaceae, Barberry Family

Prickly shrub about 3–8 feet (1–2.4 m) tall; in various arid habitats; common. LEAVES: Evergreen, trifoliate, alternate; leaflets 1–4 inches (2.5–10 cm) long; stiff, green or gray; margins with needle-sharp bristles on tips of teeth or lobes. FLOWERS: Late winter–early spring; lemon yellow; fragrant. FRUIT: Late spring; juicy red drupe, to ⅜ inch (1 cm) in diameter; good for jellies, too acidic for eating raw. RANGE: Texas to Arizona; Mexico. RELATED SPECIES: 3 species with 5 or more leaflets, all known as barberries. Two are shrubs; *Mahonia haematocarpa*, in the Trans-Pecos mountains, and *M. swaseyi*, endemic to the Edwards Plateau. *Mahonia repens,* an endangered species, is a nearly prostrate plant with blue fruit, in the Guadalupe Mountains. NOTE: The yellow wood of *Mahonia* species makes a good yellow dye.

208. River Birch, *Betula nigra*
Betulaceae, Birch Family

Small to large tree to 90 feet (27.7 m); reddish or gray bark peels in thin curled papery strips; lenticels present (short, corky horizontal lines on bark); along streams and river bottoms; uncommon. LEAVES: Deciduous, simple, alternate; blade egg-shaped to triangular or diamond-shaped, 1–4 inches (2.5–10 cm) long; bottom whitish, may be velvety; margins sharp-toothed. FLOWERS: Spring; inconspicuous catkins. FRUIT: Conelike, about 1½ inches (4 cm) long, with numerous winged nutlets that soon fall off stalk. RANGE: Eastern U.S. RELATED SPECIES: Alder (*Alnus serrulata*), in East Texas wetlands has smooth to somewhat scaly bark with lenticels, elliptic to egg-shaped leaves (widest near tip), and tiny cones with persistent woody scales.

209. Ironwood, American Hornbeam, Lechillo, *Carpinus caroliniana* Betulaceae, Birch Family

Small to medium-sized tree to 49 feet (15 m) tall; bark gray, trunk smooth, fluted, muscular-looking; in woodlands and along streams; common. LEAVES: Deciduous, simple, alternate; blade elliptic to lanceolate, 2–3½ inches (5–9 cm) long; margins sharp-toothed. FLOWERS: Spring; inconspicuous catkins. FRUIT: Tiny nutlet attached to leaflike bract; bracts form small clusters. RANGE: Southeastern U.S.

210. American Hop Hornbeam, *Ostrya virginiana* Betulaceae, Birch Family

Small to medium-sized tree to 65 feet (20 m) tall; bark shaggy; in forests and along streams; common. LEAVES: Deciduous, simple, alternate; blade egg-shaped to elliptic or lanceolate, 2–4½ inches (5–12 cm) long; often covered with soft hairs; margins with sharp teeth. FLOWERS: Spring; inconspicuous catkins. FRUIT: Tiny nutlet enclosed in inflated papery sack, about ¾ inch (2 cm) long; sacks in pendant cluster. RANGE: Eastern U.S. RELATED SPECIES: *Ostrya chisosensis*, an endangered species in the Chisos Mountains, and *O. knowltonii*, in the Guadalupe Mountains, have smaller leaves.

211. Catalpa, *Catalpa speciosa* Bignoniaceae, Catalpa Family

Medium-sized tree to 64 feet (19.7 m) tall; usually in moist woods; uncommon. LEAVES: Deciduous, simple, opposite or whorled; blade may be heart-shaped, 6–14 inches (1.5–3.5 dm) long and wide; tip long, tapering; margins entire or sometimes with shallow lobes. FLOWERS: Spring; white, showy, with yellow stripes and pale purple spots in throat; fragrant; trumpet-shaped, 1–2 inches (2.5–5 cm) long. FRUIT: Brown to black cylindrical pod 8–20 inches (2–5 dm) long; seeds flat, winged. RANGE: Eastern U.S. NOTE: Catalpa is commonly used in landscaping. Pods can be used in basket making. RELATED SPECIES: *Catalpa bignonioides* has a leaf tip that is not long and tapering; the valves of the pod are flat rather than rounded after the pod splits open.

212. Desert Willow, Mimbre, *Chilopsis linearis*
Bignoniaceae, Catalpa Family

Shrub or small tree to about 30 feet (9 m) tall; in desert washes and canyons, along roadsides; common to abundant; several varieties. LEAVES: Deciduous, simple, alternate or opposite; blade linear to narrowly lanceolate, 1–5 inches (2.5–12.5 cm), or occasionally 1 foot (3 dm) long, to ½ inch (1.2 cm) wide; surface may be sticky; margins entire. FLOWERS: Spring–fall; pink to purple, sometimes purple-striped, variegated, or white; fragrant; trumpet-shaped, 1–1½ inches (2.5–4 cm) long. FRUIT: Cylindrical pod 4–12 inches (1–3 dm) long, light brown; seeds flat, winged. RANGE: Texas to California; Mexico. NOTE: Desert willow is used ornamentally and naturalized north of its natural range. SIMILAR SPECIES: See willow (*Salix*), page 271.

213. Yellow Trumpet Flower, Retama, *Tecoma stans*
Bignoniaceae, Catalpa Family

Shrub or small tree to about 10 feet (3 m) tall (occasionally taller); in full sun, in dry rocky soil, in mountains; uncommon. LEAVES: Deciduous, pinnately compound, opposite; 5–13 leaflets, elliptic, lanceolate, or linear, 1–4 inches (2.5–10 cm) long; shiny, dark green, tip pointed; margin sharp-toothed. FLOWERS: Spring–fall; bright yellow to orange; fragrant; trumpet-shaped, 5-lobed, 1–3 inches (2.5–8 cm) long. FRUIT: Brown cigar-shaped pod 4–8 inches (1–3 dm) long; seeds flat, winged. RANGE: Texas to Arizona; Mexico. NOTE: This species is frequently cultivated in South Texas but freezes to the ground below 20°F.

214. Wild Olive, Anacahuita, *Cordia boissieri*
Boraginaceae, Borage Family

Shrub or small tree to 27 feet (8.3 m); in ravines and dry hills, along streams and roadsides; uncommon. LEAVES: Evergreen, simple, alternate; blade heart-shaped to broadly lanceolate or elliptic, 2–8 inches (5–20 cm) long; surface densely hairy, top rough, bottom velvety; margins entire or finely toothed. FLOWERS: Year-round; white with yellow or brown in throat; showy, wrinkled petals forming trumpet 1–2 inches (2.5–5 cm) long. FRUIT: Fleshy drupe about 1 inch (2.5 cm) long, shiny, white to brownish, partially enclosed within densely hairy sepals. RANGE: Texas; Mexico. NOTE: This species is frequently cultivated in South Texas but freezes back below 20°F.

215. Anacua, Sugarberry, *Ehretia anacua*
Boraginaceae, Borage Family

Small to medium-sized tree to 42 feet (12.9 m) tall, with single or multiple trunks, dense crown, and deeply furrowed bark; in moist soil, along rivers, in woods and thickets; common. LEAVES: Evergreen in mild winters, simple, alternate; blade broad, elliptic to egg-shaped, 1–5 inches (2.5–12.5 cm) long; dark green, stiff, sandpapery rough; margins entire, wavy, or with coarse teeth. FLOWERS: Spring and fall; white; aromatic, 5-lobed, ¼–½ inch (7–12 mm) wide; in small clusters. FRUIT: Yellow to red drupe ¼ inch (8 mm) in diameter; sweet edible pulp. RANGE:

Texas; Mexico. NOTE: Anacua is often planted as a shade tree but dies back in hard freezes.

216. White Honeysuckle, *Lonicera albiflora*
Caprifoliaceae, Honeysuckle Family

Arching shrub or woody vine; on rocky hills, in cedar brakes and mountains; bark shredding; common. LEAVES: Deciduous, simple, opposite; pair below flowers may be united at base; blade round to egg-shaped, 1–2½ inches (2.5–6.5 cm) long; top olive green; bottom pale, smooth to densely hairy; margins entire. FLOWERS: Spring; white to yellowish; fragrant; ¾ inch (2 cm) long, funnel-shaped, bilateral. FRUIT: Tight clusters of round orange to red berries ³⁄₁₆ inch (5 mm) wide. RANGE: Southwestern U.S.; Mexico. RELATED SPECIES: Coral honeysuckle (*Lonicera sempervirens*) is a climbing vine with red flowers (see page 136).

217. Elderberry, *Sambucus nigra* ssp. *canadensis*
Caprifoliaceae, Honeysuckle Family

Shrub to about 15 feet (4.6 m) tall; branches covered with corky bumps; along streams, in wetlands; common. LEAVES: Deciduous, pinnately compound, opposite; leaflets egg-shaped to lanceolate, 2–7 inches (5–18 cm) long; margins sharp-toothed. FLOWERS: Late spring–summer; tiny white flowers in curved or flat-topped clusters, to 15 inches (3.8 dm) in diameter. FRUIT: Tiny, juicy, blue-black, berrylike. RANGE: Eastern U.S.; Canada; many horticultural varieties grown outside natural range. NOTES: The cooked ripe fruit is edible, but raw or unripe fruit and other plant parts are somewhat toxic. RELATED SPECIES: a rare Trans-Pecos subspecies occurs in the Chisos and Davis mountains and in El Paso and Presidio counties.

218. Coralberry, *Symphoricarpos orbiculatus*
Caprifoliaceae, Honeysuckle Family

Small thicket-forming shrub to about 6 feet (2 m) tall, with delicate branches; thin bark shredding on old branches; in woods and along streams; common. LEAVES: Deciduous, simple, opposite; blade oval, egg-shaped, or nearly round, ½–2½ inches (1.2–6.5 cm) long; margins entire or slightly wavy. FLOWERS: Spring–summer; tiny; white to pink; bell-shaped, in short spikes. FRUIT: Red, pink, or purple drupes to ⅜ inch (1 cm) long, in tight clusters. RANGE: Eastern U.S. to Colorado; Mexico. NOTES: This valuable landscape plant has several horticultural varieties. RELATED SPECIES: 5 species occur in the mountains of the Trans-Pecos, all with white berries. One endemic may be endangered.

(photos 219 and 220) Viburnum, Blackhaw, Witherod,
Viburnum species (see Key)
Caprifoliaceae, Honeysuckle Family

Shrubs or small trees, predominantly in moist woods. LEAVES: Deciduous in most species, simple, opposite; blades broad, egg-shaped to elliptic to nearly round in outline; length varies among species, 1–6 inches (2.5–15 cm). FLOWERS: Spring; white; less than ⅜ inch (1 cm) wide, stamens protruding; many flowers in curved or flat-topped cluster 2–6 inches (5–15 cm) wide. FRUIT: Blue-black round to oval drupe, generally less than ½ inch (1.2 cm) long; some specimens sweet.

KEY TO *VIBURNUM* SPECIES

1a. Leaf margin with fine teeth or no teeth . . . . go to 4

1b. Leaf margin with large, coarse teeth or lobes . . . . go to 2

2a (1b). Small shrub in Trans-Pecos mountains . . . . *Viburnum rafinesquianum*

2b. Shrub or small tree in East Texas . . . . go to 3

3a (2b). Some leaves with 3 lobes, resembling maple leaf . . . .
. . . . arrowwood, mapleleaf viburnum, *Viburnum acerifolium*

194. Box elder,
     Acer negundo

193. Bigtooth maple,
     Acer
     grandidentatum

195. Red maple,
     Acer rubrum

197. Flameleaf sumac,
     Rhus lanceolata

196. Fragrant sumac,
     Rhus aromatica

198. Littleleaf sumac, *Rhus microphylla*

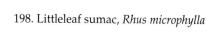

199. Evergreen sumac,
   Rhus virens

200. American holly,
   Ilex opaca

201. Yaupon holly,
   Ilex vomitoria

202. Devil's walking
   stick,
   Aralia spinosa

203. Dwarf palmetto,
   Sabal minor

204. Sabal palm,
   Sabal mexicana

205. Povertyweed,
     Baccharis neglecta

206. Tarbush,
     Flourensia cernua

207. Agarita,
     Mahonia trifoliolata

208. River Birch,
     Betula nigra

209. Ironwood,
     Carpinus caroliniana

210. American hop
     hornbeam,
     Ostrya virginiana

211. Catalpa,
Catalpa speciosa

212. Desert willow,
Chilopsis linearis

214. Wild olive, *Cordia boissieri*

213. Yellow trumpet
flower,
Tecoma stans

215. Anacua,
Ehretia anacua

216. White
honeysuckle,
Lonicera albiflora

217. Elderberry,
    Sambucus nigra
    ssp. *canadensis*

218. Coralberry,
    Symphoricarpos
    orbiculatus

219. Southern
    arrowwood
    Viburnum dentatum

220. Rusty blackhaw,
    Viburnum rufidulum

221. Four-wing saltbush,
    Atriplex canescens

222. Tumbleweed,
    Salsola tragus

223. Roughleaf dogwood, *Cornus drummondii*

224. Flowering dogwood, *Cornus florida*

225. Water tupelo, *Nyssa aquatica*

226. Arizona cypress, *Cupressus arizonica*

227. Alligator juniper, *Juniperus deppeana*

228. Red-berry juniper, *Juniperus pinchotii*

229. Eastern red cedar,
Juniperus virginiana

230. Texas persimmon,
Diospyros texana

231. Common
persimmon,
*Diospyros
virginiana*

233. Texas madrone,
Arbutus xalapensis

232. Mormon tea,
Ephedra species

234. Hoary azalea,
Rhododendron canescens

235. Farkleberry,
 *Vaccinium
 arboreum*

236. Candelilla,
 *Euphorbia
 antisyphilitica*

237. Southwest
 bernardia,
 Bernardia myricifolia

238. Leatherstem,
 Jatropha dioica

239. Chinese tallow,
 Triadica sebifera

240. Guajillo,
 Acacia berlandieri

241. Catclaw,
    Acacia greggii

242. Viscid acacia,
    *Acacia
    neovernicosa*

243. Huisache,
    Acacia farnesiana

245. Redbud, *Cercis canadensis*

244. False indigo,
    Amorpha fruticosa

246. Feather dalea,
    Dalea formosa

247. Coral bean,
  Erythrina herbacea

248. Kidneywood,
  Eysenhardtia texana

249. Honey locust,
  Gleditsia triacanthos

250. Goldenball leadtree,
  Leucaena retusa

251. *Mimosa borealis*

252. *Mimosa dysocarpa*

253. *Mimosa asperata*

254. Retama,
 Parkinsonia aculeata

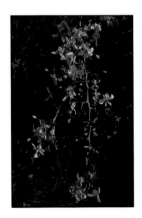

255. Paloverde,
 Parkinsonia texana

256. Texas ebony,
 Ebenopsis ebano

257. Honey mesquite,
 *Prosopis
 glandulosa*

258. Tornillo,
 Prosopis pubescens

259. Black locust,
     Robinia pseudoacacia

260. Rattlebush,
     Sesbania drummondii

261. Eve's necklace,
     Sophora affinis

262. Texas mountain
     laurel,
     Sophora secundiflora

263. Chinquapin,
     Castanea pumila

264. Beech,
     Fagus grandifolia

266. Emory oak, *Quercus emoryi*

265. Texas oak,
   Quercus buckleyi

268. Chisos red oak,
   Quercus gravesii

267. Southern red oak,
   Quercus falcata

269. Gray oak,
   Quercus grisea

270. Laçey oak,
   Quercus laceyi

271. Bur oak, *Quercus macrocarpa*

272. Blackjack
oak,
*Quercus
marilandica*

274. Water oak, *Quercus nigra*

273. Chinkapin oak,
*Quercus
muehlenbergii*

275. Willow oak,
Quercus phellos

276. Vasey oak,
Quercus vaseyana

277. Shin oak,
     Quercus sinuata
     var. *breviloba*

278. Post oak,
     Quercus stellata

279. Coastal live oak,
     *Quercus
     virginiana*

280. Ocotillo, *Fouquieria splendens*

281. Silktassel,
     Garrya ovata

282. Witch hazel,
     Hamamelis virginiana

284. Red buckeye,
Aesculus pavia
var. *pavia*

283. Sweetgum,
*Liquidambar
styraciflua*

285. St. Andrew's
Cross,
*Hypericum
hypericoides*

286. Pecan,
Carya illinoinensis

288. Little
walnut,
*Juglans
microcarpa*

287. Black hickory,
Carya texana

289. Allthorn,
Koeberlinia spinosa

290. Range ratany,
Krameria erecta

292. Red bay,
Persea borbonia

291. Shrubby blue sage,
Salvia ballotiflora

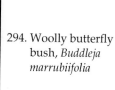

293. Sassafras,
Sassafras albidum

294. Woolly butterfly
bush, *Buddleja
marrubiifolia*

295. Southern magnolia,
     Magnolia grandiflora

296. Rose
     pavonia,
     *Pavonia
     lasiopetala*

297. Chinaberry,
     Melia azedarach

298. Paper mulberry,
     *Broussonetia
     papyrifera*

299. Osage orange,
     Maclura pomifera

300. Mulberry, *Morus* species

301. Wax myrtle,
     Morella cerifera

302. Fringe tree,
     Chionanthus
     virginicus

303. Narrow-leaf
     forestiera,
     Forestiera
     angustifolia

304. Spring herald,
     Forestiera pubescens

306. Texas ash,
     Fraxinus texensis

305. White ash,
     Fraxinus americana

307. Fragrant ash,
Fraxinus cuspidata

308. Ligustrum,
Ligustrum japonicum

309. Colorado pinyon,
Pinus edulis

310. Loblolly pine,
Pinus taeda

311. Sycamore,
Platanus occidentalis

312. Redroot,
Ceanothus americanus

313. Desert ceanothus,
    Ceanothus greggii

314. Snakewood,
    Colubrina texensis

315. Javelina bush,
    Condalia ericoides

316. Brasil,
    Condalia hookeri

317. Coyotillo,
    Karwinskia humboldtiana

318. Carolina buckthorn,
    Rhamnus caroliniana

319. Lotebush, *Ziziphus obtusifolia*

320. Mountain mahogany, *Cercocarpus montanus*

321. Hawthorn, *Crataegus texana*

322. Apache plume, *Fallugia paradoxa*

323. Laurel cherry, *Prunus caroliniana*

324. Black cherry, *Prunus serotina*

326. Scarlet bouvardia,
Bouvardia ternifolia

325. Chickasaw plum,
Prunus
angustifolia

327. Buttonbush,
Cephalanthus
occidentalis

328. Hop tree,
Ptelea trifoliata

329. Lime prickly ash,
Zanthoxylum fagara

330. Hercules'club,
Zanthoxylum
clava-herculis

331. Eastern
     cottonwood,
     Populus deltoides

332. Black willow,
     Salix nigra

333. Western
     soapberry,
     *Sapindus
     saponaria*

335. Coma,
     *Sideroxylon
     lanuginosum*

334. Mexican buckeye,
     Ungnadia speciosa

336. Cenizo,
     Leucophyllum frutescens

337. Tree of heaven,
    Ailanthus altissima

338. Tree tobacco,
    Nicotiana glauca

339. Texas snowbell,
    Styrax texanus

340. Tamarisk,
    Tamarix species

341. Bald cypress,
    Taxodium distichum

342. Basswood,
    Tilia americana

343. Hackberry,
    Celtis laevigata

344. Granjeno,
    Celtis pallida

345. Winged elm,
    Ulmus alata

346. Cedar elm,
    Ulmus crassifolia

347. Bee brush,
    Aloysia gratissima

348. American beauty berry,
    Callicarpa americana

349. Texas lantana, *Lantana urticoides*

350. Guayacán,
Guajacum angustifolium

351. Creosote bush,
Larrea tridentata

352. Texas hedgehog,
Sclerocactus
uncinatus var.
wrightii

353. Living rock,
Ariocarpus
fissuratus

354. Horse crippler,
Echinocactus
texensis

355. Strawberry cactus,
  *Echinocereus
  stramineus*

356. Rainbow cactus,
  *Echinocereus
  dasyacanthus*

357. Lace cactus,
  *Echinocereus
  reichenbachii* var.
  reichenbachii

358. Claret cup cactus,
  Echinocereus triglochidiatus

359. Nipple cactus,
  Mammillaria heyderi

360. Cane cholla, *Opuntia imbricata*

361. Tasajillo,
     Opuntia leptocaulis

362. Texas prickly pear,
     Opuntia engelmannii

363. Dog cholla,
     Opuntia schottii

364. Purple prickly pear,
     Opuntia macrocentra

365. *Agave havardiana*

366. Lechuguilla,
     Agave lechuguilla

367. Sotol,
Dasylirion species

368. Red yucca,
Hesperaloë
parviflora

369. Beargrass, *Nolina erumpens*

370. Soaptree yucca,
Yucca elata

371. Pale-leaf yucca,
Yucca pallida

372. Twisted-leaf
yucca,
Yucca rupicola

373. Trecul yucca,
  Yucca treculeana

374. Hechtia,
  Hechtia texensis

375. Giant ragweed,
  Ambrosia trifida

376. Cocklebur,
  Xanthium
  strumarium

377. Carrizo,
  Arundo donax

378. Cattail,
  Typha species

380. Ball moss, *Tillandsia recurvata*

379. Gall on live oak

381. Spanish moss,
Tillandsia usneoides

382. Mistletoe,
*Phoradendron
tomentosum*

383. Lichens on tree

384. Moss

3b. Leaves not lobed . . . .
. . . . southern arrowwood, *Viburnum dentatum* (photo 219)

4a (1a). Leaf margins entire or wavy (occasionally with fine teeth); only in East Texas; leaf bottom and leaf stem may bear rusty red scales, . . .
. . . . possum haw, *Viburnum nudum*

4b. Leaf margins with fine teeth; in East, Central, or Trans-Pecos Texas . . . .
. . . . go to 5

5a (4b). Top surface of leaf shiny; bottom of leaf or leafstalk usually with rusty red hairs or scales; common in East and Central Texas, with a population in Davis Mountains; shrub or small tree with checkered bark . . . . rusty blackhaw, *Viburnum rufidulum* (photo 220)

5b. Top surface of leaf dull green; leaf bottom and stalk lack reddish hairs; uncommon to rare, in East Texas . . . .
. . . . blackhaw, *Viburnum prunifolium*

221. Four-wing Saltbush, *Atriplex canescens*
Chenopodiaceae, Goosefoot Family

Small shrub to about 6 feet (2 m) tall; in dry rocky, sandy, or saline soil; widespread and abundant. LEAVES: Evergreen, simple, alternate; blade oblong to linear, folding inward, ⅜–2 inches (1–5 cm) long; foliage gray-green, fleshy, aromatic; margins entire. FLOWERS: Spring–fall; tiny, lacking petals, covered with whitish hairs; in small spikes; dioecious; abundant pollen makes male shrubs rusty yellow. FRUIT: Seed 4-winged, papery when dry; in dense showy spikes. RANGE: Texas and Mexico to California, north to Canada. RELATED SPECIES: This is the most widespread of 17 species.

222. Tumbleweed, Russian Thistle,
Salsola tragus (S. iberica, S. kali)
Chenopodiaceae, Goosefoot Family

Annual shrubby spiny weed with rounded shape, about 2–9 feet (6–30 dm) tall; branches pencil-size, green, streaked with red or purple, turning brown in fall when plant dies and is blown by wind; in dry disturbed areas, usually in sandy soil, along roadsides, in fields and plains; abundant in some areas. LEAVES: Spiny, succulent, or threadlike; spine-tipped; minute (6 mm) or to 3 inches (7.5 cm) long. FLOWERS: Inconspicuous. FRUIT: Tiny, flat, round winged seed. RANGE: Native of Eurasia, now naturalized throughout western U.S. and Canada; found in West, South, and Coastal Texas.

223. Roughleaf Dogwood, *Cornus drummondii*
Cornaceae, Dogwood Family

Shrub or small tree to 28 feet (8.6 m) tall, with gray to reddish bark; usually in moist woods and thickets, along streams, occasionally in dry rocky soil; common. LEAVES: Deciduous, simple, opposite; blade oval to egg-shaped, 2–5 inches (5–12.5 cm) long; surface smooth or slightly rough; veins parallel, conspicuous; margins entire. FLOWERS: Spring–summer; white; tiny; in showy flat-topped or rounded clusters 2–3 inches (5–7.5 cm) wide. FRUIT: White drupe, ¼ inch (6 mm) in diameter; may be somewhat toxic. RANGE: Southeastern U.S. RELATED SPECIES: Two East Texas species; *Cornus foemina*, with blue fruit, and *C. racemosa*, with elongated flower clusters and white fruit.

224. Flowering Dogwood, *Cornus florida*
Cornaceae, Dogwood Family

Large shrub or small tree to 37 feet (11.4 m); understory plant with spreading crown; in moist woods, in sandy soil; common. LEAVES: Deciduous, simple, opposite; blade oval to egg-shaped, 3–6 inches (7.5–15 cm) long; turning red in fall; veins parallel, conspicuous; margins entire. FLOWERS: Early spring; 4 showy white or pink bracts surround tight clusters of minute greenish flowers. FRUIT: Tight clusters of red drupes; may be somewhat toxic. RANGE: Eastern U.S. NOTE: A popular ornamental, flowering dogwood has numerous horticultural varieties, including a red-flowered form. Branches can be used in basket making. Leaves and woody parts yield nice dyes.

225. Water Tupelo, *Nyssa aquatica*
Cornaceae, Dogwood Family

Large semiaquatic tree usually with swollen base, to more than 100 feet (30 m) tall; in swamps, along bayous; common. LEAVES: Deciduous, simple, alternate or crowded at branch tips; blade broad, elliptic to egg-shaped, 3–12 inches (8–30 cm) long; bottom whitish, often covered with soft fuzzy hairs; margins entire or with several prominent teeth, tapering tip. FLOWERS: Spring, emerging with or before leaves; minute; inconspicuous. FRUIT: Oblong purple drupe about 1 inch (2.5 cm) long. RANGE: Eastern U.S. RELATED SPECIES: Black tupelo (*Nyssa sylvatica*), found in swamps and sandy bottomlands, has broad leaves less than 6 inches (1.5 dm) long, with top shiny and bottom of mature leaves smooth (if has hairs, not fuzzy), and fruit to ⅝ inch (1.5 cm) long; the trunk lacks the widened base of the other two species. *Nyssa biflora* has narrower, leathery leaves.

226. Arizona Cypress,
Cedro Blanco, *Cupressus arizonica*
Cupressaceae, Cypress Family

Small to large coniferous tree to 112 feet (34 m) tall; bark of small trees sheds, exposing smooth red bark or gray bark; in Chisos Mountains; uncommon. LEAVES: Evergreen, minute triangular scales; gray-green; strong-smelling, may be sticky. FLOWERS: In tiny cones. FRUIT: Woody cone 1 inch (2.5 cm) in diameter, separating into 6–8 small shields. RANGE: Southwestern U.S.; Mexico. NOTE: Often used in landscapes, Arizona cypress has several horticultural varieties. It favors hot dry summers, cold winters, a sunny site and well-draining soil.

(photos 227, 228, 229) Juniper, Cedar, Sabino,
Juniperus species (see Key)
Cupressaceae, Cypress Family

Evergreen coniferous shrubs or trees. LEAVES: Scalelike or needlelike; minute, surrounding branchlets; aromatic. FLOWERS: Minute cones; most species dioecious; pollen of some species highly allergenic. FRUIT: Small juicy or woody berrylike cones; in most species, cones to ½ inch (1.2 cm) in diameter, coated with a whitish wax (a bloom). NOTE: Hybrids cause difficulty in identification. Various non-native conifers are also known as cedars, such as the ornamental deodar cedar (*Cedrus deodara*).

KEY TO JUNIPERS

1a. From Dallas, Austin, and Bastrop west . . . . go to 3

1b. In East Texas (west to Bastrop, Austin, and Dallas) . . . . go to 2

2a (1b). Common and widespread; branch tips weak, may be drooping, but branches generally horizontal or turning upward . . . .
Eastern red cedar, *Juniperus virginiana* (photo 229)—Medium to large trees, often pyramidal or columnar in shape; bark shredding; mature cones blue, ³⁄₁₆–⁵⁄₁₆ inch (5–8 mm) in diameter; northwestern edge of range extends to Wichita Falls, with isolated populations in northern Panhandle; overlapping with ranges of several western junipers.

2b. East Texas, mainly near the coast, with some populations farther inland; uncommon; branches drooping . . . .
Southern red cedar, *Juniperus virginiana* var. *silicicola*—Medium-sized tree, crown pyramidal or rounded; bark shredding; mature cones blue, to ³⁄₁₆ inch (5 mm) in diameter.

3a (1a). In the Trans-Pecos . . . . go to 6

3b. In central Texas or the Panhandle, from Austin west to the Pecos River
. . . . go to 4

4a (3b). Fruit copper to reddish brown, lacking a white waxy bloom . . . .
Red-berry juniper, *Juniperus pinchotii* (photo 228)—Large shrub or small tree, common to abundant throughout West Texas; in the Trans-Pecos mountains, most common below 5500 feet elevation; cones ¼–⅜ inch (6–10 mm) in diameter, with 1 or 2 seeds; branch tips stiff, usually erect; allergenic pollen in fall; hybridizes with several other species.

4b. Fruit blue, usually with a white waxy bloom . . . . go to 5

5a (4b). Abundant in Central Texas, extending west to edge of the Trans-Pecos, northeast to Dallas, and south to Zavala County; cones blue, ¼–⁵⁄₁₆ inch (6–8.5 mm) long, with 1 or 2 seeds . . . .
Mountain cedar, ashe juniper, *Juniperus ashei*—large shrub to fairly large tree; bark shredding; often with whitish fungi on trunk; allergenic pollen in winter; range overlaps with range of eastern red cedar (2a).

5b. In the Panhandle; uncommon . . . . one of the following:

(i) One-seeded juniper, *Juniperus monosperma*—large shrub or small tree in Panhandle, also in mountains of Culberson County, possibly in Reeves County; branch tips usually stiff, erect; cones blue, ⅙–¼ inch (4–7 mm) long, with 1 seed (rarely 2 or 3).

(ii) Rocky mountain juniper, *Juniperus scopulorum*—Shrub to fairly large tree, in Panhandle and in Guadalupe Mountains, above 6000 feet in elevation; cones blue, ³⁄₁₆–⁵⁄₁₆ inch (5.5–8.5 mm) in diameter, usually with 1 or 2 seeds; branch tips erect or drooping.

6a (3a). Branches drooping, as though wilted or weeping; bark shredding or furrowed; a small to medium tree; in Texas, found only in Chisos Mountains where it is common above 5000 feet; with a bloom, the green to reddish brown cones appear bluish; cones ⁵⁄₁₆–½ inch (8.5–12 mm) in diameter, with 4–13 seeds . . . .
. . . . weeping juniper, drooping juniper, *Juniperus flaccida*

6b. Not as above . . . . go to 7

7a (6b). Bark deeply furrowed and checkered with rectangular scales, like alligator skin; branch tips usually stiff, erect; with a bloom, green to reddish brown cones appear bluish; cones ⁵⁄₁₆–½ inch (8.5–12 mm) in diameter, with 1–6 seeds; small to large tree in Trans-Pecos mountains, abundant above 5500 feet . . . .
alligator juniper, *Juniperus deppeana* (photo 227) (a very rare variety in Davis and Guadalupe Mountains, has branches somewhat drooping and bark furrowed and shredding).

7b. Not as above; fruit usually less than ⁵⁄₁₆ inch (8.5 mm) in diameter, rarely to ⅜ inch (1 cm) . . . . go to 8

8a (7b). Fruit copper to reddish brown, lacking white waxy bloom; common throughout west Texas . . . .
. . . . red-berry juniper, *Juniperus pinchotti* (refer to 4a)

8b. Mature fruit blue, blue-black, pink, rose, or purple, with white waxy bloom; uncommon . . . . one of the following:

(i) Rose-fruited juniper, *Juniperus erythrocarpa*—in southern and western Trans-Pecos in the mountains mainly between 4000 and 5500 feet in elevation; cones pink, rose, or purplish, with 1 or 2 seeds.

(ii) One-seeded juniper, *Juniperus monosperma* (see 5b)

(iii) Rocky mountain juniper, *Juniperus scopulorum* (see 5b)

230. Texas Persimmon, Chapote, *Diospyros texana*
Ebenaceae, Ebony Family

Shrub or small tree, 6–26 feet (2–8 m) tall; thin bark peels off to leave smooth gray trunk; in dry rocky brush or woods; common. LEAVES: Deciduous or persistent, simple, alternate; blade egg-shaped (widest at tip) to oblong, 1–2½ inches (2.5–6.5 cm) long; bottom hairy; margins entire, rolling under slightly. FLOWERS: Late winter–spring; white or creamy; fragrant; bell-shaped, to ½ inch (1.2 cm) long; dioecious. FRUIT: Late summer–fall; black, juicy, 1 inch (2.5 cm) wide, with several large seeds; edible when fully ripe, astringent when green; important wildlife food. RANGE: Texas; Mexico.

231. Common Persimmon, *Diospyros virginiana*
Ebenaceae, Ebony Family

Medium-sized tree to 60 feet (18.5 m) tall; bark furrowed; in many soil types, in dry woods, fields; common. LEAVES: Deciduous, simple, alternate; blade elliptic, oblong, or egg-shaped, 2–6 inches (5–15 cm) long; top dark green, shiny; margins entire. FLOWERS: Spring–summer; yellowish; female flowers bell-shaped, to ¾ inch (2 cm) long, males minute, inconspicuous, dioecious. FRUIT: Yellow, orange, or red; ¾–2½ inches (2–6.5 cm) in diameter, with several large seeds; sweet and edible when fully ripe, astringent when green. RANGE: Eastern U.S. NOTE: This is a small-fruited relative of the Japanese persimmon found in groceries.

232. Mormon Tea, Cañatilla,
Joint Fir, *Ephedra* **species**
Ephedraceae, Ephedra Family

Small shrubs, erect or vining; with stiff straight branches; in some species, branches vertical and broomlike; terminal branches jointed, cylindrical, to ⅛ inch (4 mm) in diameter, green or yellow; in arid, gravely, sandy, or rocky soil, deserts; uncommon. LEAVES: Deciduous or persistent, simple, opposite or in whorls of 3; scalelike, minute, with inconspicuous membranes. FLOWERS AND FRUIT: In tiny cones to ⅝ inch (1.5 cm) long, with fleshy or papery scales, emerging at joints; usually dioecious. RANGE: South Texas and western two thirds of Texas. RELATED SPECIES: Six species in Texas. NOTE: A tea can be made from the young branches. A Chinese species is high in ephedrine, used for treating hay fever.

233. Texas Madrone, Madroño,
Manzanita, *Arbutus xalapensis* (*A. texana*)
Ericaceae, Heath Family

Evergreen tree to 32 feet (9.8 m) tall, with smooth white to red bark, peeling in sheets; in scattered populations, on rocky hills, in canyons and mountains; rare. LEAVES: Evergreen, simple, alternate, but may appear whorled at branch tips; blade elliptic or oblong, to egg-shaped, 1–4 inches (2.5–10 cm) long; margins toothed or entire. FLOWERS: Late winter–spring; white to pinkish; ¼ inch (7 mm) long, bell-shaped, 5-lobed. FRUIT: Red fleshy berry to ⅜ inch (1 cm) wide, bumpy; edible. RANGE: Texas and New Mexico to Guatemala. RELATED SPECIES: Mexican manzanita (*Arctostaphylos pungens*), a rare shrub with a red trunk, grows in the Davis Mountains.

234. Hoary Azalea, *Rhododendron canescens*
Ericaceae, Heath Family

Small shrub about 3–9 feet (1–3 m) tall; in sandy, acid soil, in wetland forests, along bogs, streams, and seeps; uncommon. LEAVES: Deciduous, simple, alternate or clustered at branch tips; blade narrow, 1½–4¼ inches (4–11 cm) long; hairy; margins entire or very finely toothed. FLOWERS: In spring, before or with emerging leaves; pink (sometimes white); sticky; 5-lobed trumpet about 1 inch (2.5 cm) wide, with slender tube; stamens protruding well beyond petals. FRUIT: Elliptical capsule ⅝ inch (1.5 cm) long. RANGE: Southeastern U.S. RELATED SPECIES: 3 similar species in East Texas. NOTE: All parts, including flowers, are poisonous.

235. Farkleberry, Sparkleberry, *Vaccinium arboreum*
Ericaceae, Heath Family

Shrub or small tree to 29 feet (8.9 m) tall; understory shrub in dry sandy soil in pine and hardwood forests and fields; common. LEAVES: Persistent, simple, alternate; blades elliptic to egg-shaped (widest at tip); ½–3 inches (1.2–7.5 cm) long; often shiny, leathery; margins entire (rarely with few small teeth). FLOWERS: Spring; creamy white or pink; fragrant; tiny, to ¼ inch (6 mm) long, 5-lobed, urn-shaped; in leafy racemes. FRUIT: Reddish to black berry ¼–⅜ inch (6–10 mm) in diameter; edible but fairly dry; good for jellies. RANGE: Eastern U.S. RELATED SPECIES: Blueberries are *Vaccinium* species.

236. Candelilla, *Euphorbia antisyphilitica*
Euphorbiaceae, Spurge Family

Clump-forming shrub to about 3 feet tall (1 m), consisting of many vertical rodlike stems, to ³⁄₁₆ inch (5 mm) in diameter, grayish green, coated with wax; sap milky; in desert; locally abundant. LEAVES: Usually absent, appearing briefly on new growth; minute, linear. FLOWERS: Spring–fall, after rains; white, pink to purple; flowerlike cup, ³⁄₁₆ inch (5 mm) wide, holds numerous male flowers surrounding single female flower. FRUIT: Tiny 3-lobed capsule, green or purple. RANGE: Chihuahuan Desert of Texas, New Mexico, and Mexico. NOTE: Candelilla wax, harvested in Mexico, is used in the United States in chewing gum, cosmetics, and floor wax. Unregulated harvesting eliminated many Texas stands in the early 1900s.

237. Southwest Bernardia, Mouse Ears, Oreja de Ratón, *Bernardia myricifolia* Euphorbiaceae, Spurge Family

Shrub about 3–10 feet (1–3 m) tall; in calcareous or sandy soil, in dry brush; common. LEAVES: Deciduous or persistent in mild winters, simple, alternate; blade variable, elliptic or lanceolate, ³⁄₈–2 inches (9–50 mm) long; top rough; bottom velvety, pale or whitish; margins round-toothed, scalloped. FLOWERS: Spring; green; minute, inconspicuous; dioecious. FRUIT: 3-lobed capsule ¼–½ inch (7–12 mm) wide. RANGE: Texas; Mexico. RELATED SPECIES: *Bernardia obovata*, usually a much smaller shrub, is widespread in the Trans-Pecos.

238. Leatherstem, Rubber Plant, Sangre de Drago, *Jatropha dioica* Euphorbiaceae, Spurge Family

Unusual fleshy low shrub about 1–3 feet (3–9 dm) tall; with flexible stems, reddish or gray, wandlike, erect; often in colonies, in arid brush and desert; common in some areas. LEAVES: Absent in dry seasons; simple, alternate or crowded on stubby branches; blade linear to very broad, about 1–2 inches (2.5–5 cm) long; margins entire or 2–3 lobed. FLOWERS: Spring–summer; white to pink; tiny, bell-shaped; sepals reddish; dioecious. FRUIT: Small capsule. RANGE: Texas; Mexico. NOTE: The clear sap turns red when exposed to air. RELATED SPECIES: Two herbaceous species with leaves 5- to 9-lobed.

239. Chinese Tallow, *Triadica sebifera*
   (Sapium sebiferum)
   Euphorbiaceae, Spurge Family

Medium-sized tree to 51 feet (15.7 m) with milky sap; common. LEAVES: Deciduous, simple, alternate; blade triangular or diamond-shaped, 1–3½ inches (2.5–9 cm) long, tip tapering; turning sunset colors in fall; margins entire. FLOWERS: Spring; in yellow catkins. FRUIT: 3-lobed capsule, with 3 hard white seeds. RANGE: Chinese ornamental escapes from cultivation in eastern half of the state; abundant around Houston and Beaumont. NOTE: The fruit and leaves are poisonous. The seeds yield industrial oil and wax.

(photos 240 through 243) Acacia,
   Acacia species **(see Key)**
   Fabaceae, Legume Family

In Texas, small to medium-sized shrubs or trees, usually thorny. LEAVES: Deciduous, twice pinnately compound, alternate; 1 or more pairs of pinnae; leaflets tiny, ⅟₁₆–⅝ inch (2–15 mm) long, mostly oval to oblong or linear; margins entire. FLOWERS: Spring or summer; yellow to white colors caused by protruding stamens, fragrant; flowers in spherical heads (resembling puffballs) or cylindrical spikes (resembling bottle brushes). FRUIT: Bean pod without spines. NOTE: The flowers are a source of honey. The bark and pods of various species are used for tannins, dyes, and gum arabic. SIMILAR SPECIES: Mimosa (page 230) and mesquite (page 234).

KEYS TO *ACACIA* SPECIES

NOTE: Identification is difficult without flowers or pods. Look for old pods on the ground. Hybrids can cause difficulty in identification.

1a. Flowers form spherical heads, resembling puffballs to about ½ inch (1.2 cm) in diameter .... go to 3

1b. Flowers form cylindrical spikes, resembling bottle brushes about ½–2½ inches (1.2–6.5 cm) long .... go to 2

2a (1b). Twigs have paired straight spines ....
   Blackbrush, chaparro prieto, *Acacia rigidula*—Shrub about 3–15 feet (1–4.5 m) tall; often forms impenetrable thickets in South Texas, west to Brewster County; leaves with 1 pair of pinnae; 2–8 leaflets per pinna; leaflets ¼–⅝ inch (8–15 mm) long, shiny, dark green; flowers in late winter to spring, creamy white to pale yellow, appearing before and as leaves emerge; fruit a dark brown to black pod, 2–5 inches (5–13 cm) long, to ¼ inch (7 mm) wide.

2b. Twigs have curved spines resembling cat claws, or spines absent; leaflets mostly ⅛–½ inch (3–12 mm) long; pod often twisted ....
   Catclaw, uña de gato, *Acacia greggii* (photo 241)—Thicket-forming shrub or small tree 3–38 feet (1–11.7 m) tall, in Trans-Pecos, South, and Central Texas, northeast to Young County; leaves with 1–3 pairs of pinnae, 3–10 pairs of leaflets per pinna; flowers in spring and summer, creamy white to pale yellow; fruit a thin pod 2–6 inches (5–15 cm) long, to 1 inch (2.5 cm) wide (*A. wrightii* now considered a variety.)

3a (1a). Flowers white to pale yellow; spines present or absent ..... go to 8

3b. Flowers bright yellow; spines straight and paired, or absent .... go to 4

4a (3b). Pods thin, somewhat flattened, may be constricted between seeds, light brown to reddish brown .... go to 6

4b. Pods plump to cylindrical, reddish-brown to black .... go to 5

5a (4b). Pods usually less than 3¼ inches (8 cm) long; surface smooth, not hairy ....
   Huisache, *Acacia farnesiana* (*A. smallii*) (photo 243)—Shrub or tree 6–48 feet (2–14.6 m) tall, in South, East, and West Texas south of Travis County (possibly north of McLennan County); leaves fernlike, with 2–8 pairs of pinnae, 10–25 pairs of leaflets per pinna; leaflets to 3/16 inch (5 mm) long; flowers in late winter to spring, bright yellow; often planted outside its natural range.

5b. Pods to 5 inches (1.3 dm) long; covered with velvety or stiff hairs; shrub or small tree about 3–20 feet (1–6 m) tall, in South Texas and Edwards Plateau . . . . . Huisachillo, *Acacia schaffneri* var. *bravoensis*

6a (4a). Leaflets threadlike; leaves usually have only 1 pair of pinnae . . . . . . *Acacia schottii*—Shrub to about 6 feet (2 m) tall, with or without spines, in southern Trans-Pecos; hybridizes with *A. constricta* and *A. neovernicosa*.

6b. Leaflets linear or oblong but not threadlike, often sticky; leaves with 1 or more pairs of pinnae . . . . go to 7

7a (6b). Leaves usually have 4–7 pairs of pinnae (leaves with 1 pair and as many as 9 pairs of pinnae may occur on 1 shrub) . . . . Whitethorn acacia, largancillo, *Acacia constricta*—Shrub or small tree about 3–18 feet (1–6 m) tall, with or without thorns, in Trans-Pecos and southern High Plains and along Rio Grande in far South Texas; leaves with 4–16 pairs of leaflets per pinna; leaflets ⅟₁₆–¼ inch (2–6 mm) long; flowers in spring to summer, bright yellow; fruit a pod 1–5½ inches (2.5–14 cm) long, to ¼ inch (7 mm) wide.

7b. Leaves with 1–3 pairs of pinnae . . . . Viscid acacia, *Acacia neovernicosa* (photo 242)—Shrub about 3–6 feet (1–2 m) tall, western Edwards Plateau to Trans-Pecos; not always possible to distinguish from *A. constricta* or *A. schottii* because of hybridization.

8a (3a). Leaves fernlike, with 3–12 (or more) pairs of pinnae . . . . . . go to 10

8b. Leaves with 1–4 pairs of pinnae . . . . go to 9

9a (8b). Shrub 3–17 feet (1–5.2 m) tall, with curved spines resembling cat claws . . . . Catclaw, *Acacia roemeriana*—In Central Texas from Travis County west to Trans-Pecos; leaflets ⅛–½ inch (3–12 mm) long; up to 8 pairs of leaflets per pinna; flowers in late winter to spring, creamy white to pale yellow; fruit a thin pod 2–4 inches (5–10 cm) long, to 1¼ inch (3 cm) wide; hybridizes with *A. berlandieri*.

9b. Low shrub usually less than 3 feet (1 m) tall, with no spines . . . . *Acacia angustissima* var. *chisosiana*—In south Brewster and Presidio counties; leaflets less than ¼ inch (6 mm) long; 6–10 pairs of leaflets per pinna; flowers in spring to summer, white.

10a (8a). Shrub or small tree to about 15 feet (4.5 m) tall, with or without small spines; flowers in spring to summer, white to pale yellow; mature pods woody, often velvety, 4–6 inches (1–1.5 dm) long, to 1¼ inch (3 cm) wide . . . .

Gaujillo, *Acacia berlandieri* (photo 240)—From South Texas, west to Trans-Pecos near the Rio Grande; when present, spines curved to nearly straight but not paired at nodes; leaflets to ¼ inch (6 mm) long; usually 20–50 pairs of tiny leaflets per pinna; hybridizes with *A. roemeriana*.

10b. Low-growing, shrubby to herbaceous, usually less than 3 feet (1 m) tall, with no spines; flowers white to pale yellow; pod papery, to 3 inches (7.5 cm) long, to ½ inch (1.2 cm) wide ....go to 11

11a (10b). Leaves with 3–6 pairs of pinnae; 9–15 pairs of leaflets per pinna .... Prairie acacia, *Acacia angustissima* var. *texensis* (*A. texensis*)—In South Texas and Trans-Pecos; leaflets less than ¼ inch (6 mm) long; flowers in spring to summer.

11b. Leaves with 7 to many pairs of pinnae; many pairs of leaflets per pinna Fern acacia, *Acacia angustissima* var. *hirta*—In eastern two thirds of Texas; leaflets less than ¼ inch (6 mm) long; flowers in spring to summer.

244. False Indigo, *Amorpha fruticosa*
Fabaceae, Legume Family

Small shrub about 3–10 feet (1–3 m) tall; often in limestone soils, in moist woods or along streams; widespread but uncommon, with 4 varieties. LEAVES: Deciduous, pinnately compound, alternate; 7–35 leaflets; leaflets oblong to elliptic, ½–2 inches (1.2-5 cm) long; ¼–¾ inch (6–20 mm) wide; margins entire. FLOWERS: Spring; purple; aromatic, in spike 4–8 inches (1–2 dm) long, resembling a bottle brush; protruding yellow-tipped stamens. FRUIT: Pod ¼ inch (7 mm) long. NOTE: This is a highly variable species with white-flowering and dwarf forms, some with variegated leaves. RANGE: U.S.; Mexico. RELATED SPECIES: 4 other species with restricted ranges.

245. Redbud, *Cercis canadensis*
Fabaceae, Legume Family

Small to medium-sized tree to 37 feet (11.4 m) tall; in woods, canyons, mountains; widespread and common, with 3 varieties.

LEAVES: Deciduous, simple, alternate; blade more or less heart-shaped, 2–6 inches (5–15 cm) long and wide; margins entire. FLOWERS: Early spring, before leaves emerge; pink to purple; pea-type, about ½ inch (1.2 cm) long. FRUIT: Thin reddish brown bean pod 2–4 inches (5–10 cm) long. RANGE: Eastern U.S., Canada, and Mexico. NOTE: Several horticultural varieties exist, including 1 with white flowers. The flowers are sweet and edible and yield nice dyes.

246. Feather Dalea, *Dalea formosa*
Fabaceae, Legume Family

Intricately branching shrub to about 4 feet (1.2 m) tall; in dry hills, desert thickets; common in Trans-Pecos. LEAVES: Deciduous, pinnately compound, alternate; 7–13 leaflets, to ⅛ inch (3 mm) long; grayish; margins entire, with edges rolled under. FLOWERS: Spring–fall; purple (upper petals may be yellow); about ½ inch (1.2 cm) long, pea-type; lobes of sepals form feathery threads ⅜ inch (1 cm) long. FRUIT: Minute pod enclosed in feathery sepals. RANGE: Southwestern U.S.; Mexico. RELATED SPECIES: 33 other species grow in Texas, mostly herbaceous wildflowers. Black dalea (*Dalea frutescens*), a shrub in the western half of Texas, lacks the feathery sepals found on feather dalea and has 9–17 leaflets, each ⅛–⅝ inch (3–8 mm) long. Other shrubby species have hairy leaves and branches.

247. Coral Bean, Colorín, *Erythrina herbacea*
Fabaceae, Legume Family

Shrub about 2–6 feet (6–20 dm) tall, dying back in cold winters; young branches smooth, green; often bearing curved thorns on branches and leafstalks; in sandy woods; uncommon. LEAVES: Deciduous, alternate, compound, with 3 leaflets per leaf; leaflets triangular to 3-lobed, 1–6 inches (2.5–15 cm) long and often as wide;

margins not toothed. FLOWERS: Spring; scarlet; 1–2 inches (2.5–5 cm) long, petals closed, appearing tubular but upper petal wrapped around lower 4 petals. FRUIT: Pod 2–8 inches (5–20 cm) long, green to black, constricted between seeds; seeds hard, scarlet, to ½ inch (1.2 cm) long, poisonous. RANGE: Southeastern U.S. to Texas and Mexico. NOTE: Coral bean is often used in landscaping.

248. Kidneywood,
   Vara Dulce, *Eysenhardtia texana*
   Fabaceae, Legume Family

   Intricately branched shrub about 3–12 feet (1–3.5 m) tall; on dry rocky or sandy soil; common. LEAVES: Deciduous, alternate, pinnately compound, with many leaflets; leaflets oblong, ⅛–½ inch (4–12 mm) long; strongly aromatic; margins entire. FLOWERS: Spring to fall; white; fragrant; tiny; in clusters on branch tips. FRUIT: Tiny pod, to ⅜ inch (1 cm) long, with 1 mature seed. RANGE: Texas; Mexico. RELATED SPECIES: *Eysenhardtia spinosa*, a smaller shrub with somewhat spiny branches, occurs in Presidio County and is in danger of extinction.

249. Honey Locust, *Gleditsia triacanthos*
   Fabaceae, Legume Family

   Shrub or large tree to 93 feet (28.6 m) tall; trunk and branches bear long, often branching thorns; in woods, along fencerows; uncommon. LEAVES: Deciduous, alternate or clustered, once or twice pinnately compound, with many leaflets; leaflets oblong to lanceolate, ½–2 inches (1.2–5 cm) long; margins entire or finely toothed. FLOWERS: Spring; green, fragrant; small, in hanging clusters. FRUIT: Large pod, often twisted, 8–18 inches (2–4.5 dm) long, dark reddish brown; seeds hard, shiny. RANGE: Eastern U.S. NOTE: Many horticultural varieties are used outside the natural range, including thornless forms. RELATED SPECIES: Water locust

(*Gleditsia aquatica*) grows in swamps and wetlands and has a small pod 1–3 inches (2.5–7.5 cm) long. Hybrids occur.

250. Goldenball Leadtree, *Leucaena retusa*
   Fabaceae, Legume Family

Shrub or small tree to 21 feet (6.5 m) tall; in dry rocky hills, canyons, mountains; uncommon. LEAVES: Deciduous, alternate, twice pinnately compound, with 2–4 pairs of pinnae; 3–9 pairs of leaflets per pinna; leaflets oval, ¼–1 inch (7–25 mm) long; margins entire. FLOWERS: Spring–fall; bright yellow; aromatic; in dense spherical heads, to 1 inch (2.5 cm) in diameter; protruding stamens produce puffball effect. FRUIT: Thin pod 4–10 inches (1–2.5 dm) long. RANGE: Texas, New Mexico; Mexico. RELATED SPECIES: Tepeguaje (*Leucaena pulverulenta*) and popinac (*L. leucocephala*), have creamy white flowers and numerous pinnae with many tiny leaflets. Tepeguaje, with linear leaflets, about ¹⁄₁₆ inch (1 mm) wide, is native to extreme South Texas and planted farther north. Popinac, not a native, has leaflets to ⅛ inch (4 mm) wide and is planted in South Texas.

(photos 251, 252, 253) Mimosa, Catclaw,
   Mimosa species **(see Key)**
   Fabaceae, Legume Family

In Texas small shrubs, vines or crawling perennials, with twigs and stems armed with curved (some with straight) spines; leaves and pods may bear spines. LEAVES: Deciduous, alternate, twice pinnately compound, with 1 to many pairs of pinnae; leaflets ¹⁄₁₆–½ inch (1–12 mm) long; margins entire; leaflets of some species fold up when touched. FLOWERS: Spring–fall; pink to purple, white in some species; fragrant; flowers form small spherical heads (cylindrical in a few species), with protruding stamens producing puffball effect. FRUIT: Small bean pods, often with spines along margins, a few species with spines on flat surface of pod. SIMILAR SPECIES: Acacias (page 224) and mesquites (page 234).

KEY TO *MIMOSA* SPECIES

1a. In the Trans-Pecos .... go to 7

1b. East of the Trans-Pecos .... go to 2

2a (1b). Erect shrubs ....go to 4

2b. Prostrate plant or climbing vine ....go to 3

3a (2b). Prostrate herbaceous plants ....
>   *Mimosa* species—6 species, difficult to distinguish, stems cov-
>   ered with stiff hairs or prickles; flowers pink to purple, globes;
>   fruits linear or oblong, bristly. See p. 35

3b. Climbing vine ....
>   *Mimosa malacophylla*—Rare, in South Texas; leaflets to ½ inch
>   (1.2 cm) long; flowers white.

4a (2a). Along southern coast, from Calhoun County to Cameron and
>   Hidalgo counties ....
>   *Mimosa asperata* (*M. pigra* var. *berlandieri*) (photo 253)—Shrub
>   about 3–9 feet (1–3 m) tall; 4–12 pairs of pinnae; often 20 or
>   more pairs of leaflets per pinna; leaflets fold up when touched;
>   leafstalk and twigs hairy; spines straight or slightly curved;
>   pods covered with stiff hairs; mature pods break into seg-
>   ments.

4b. Found elsewhere .... go to 5

5a (4b). Leaves with 3–10 pairs of pinnae ....
>   *Mimosa aculeaticarpa* var. *biuncifera*—Shrub about 3 feet (1 m)
>   tall, abundant and widespread, in Central and West Texas and
>   Panhandle; 5–14 pairs of leaflets per pinna; leaflets ¹⁄₁₆–¼ inch
>   (2–6 mm) long; flowers pink or white; pods with or without
>   spines, not breaking into segments.

5b. Leaves with 1–3 (rarely 4) pairs of pinnae, shrub about 2-6 feet (6-20
>   cm) tall .... go to 6

6a (5b). Pods yellow or light to dark brown; margins of pods with or with-
>   out prickles; pods often constricted between seeds, twisting and
>   eventually breaking into segments ....
>   *Mimosa borealis* (photo 251)—common and widespread, in Cen-
>   tral and West Texas and the Panhandle; leaflets ¹⁄₁₆–¼ inch (2–7
>   mm) long; 3–8 pairs of leaflets per pinna; flowers pink or white.

6b. Pods dark brown, with or without prickles; pods usually straight, not
>   breaking into segments ....

Mimosa texana (*M. wherryana*) on Edwards Plateau, in Trans-Pecos and South Texas; to 7 pairs of leaflets per pinna; flowers pink or white.

7a (1a). Flower heads cylindrical . . . .
Mimosa dysocarpa var. *dysocarpa* (photo 252)—Shrub about 3–6 feet (1–2 m) tall; in mountains of Brewster, Presidio, and Jeff Davis counties; leaves with numerous pairs of pinnae; leaves, twigs, and pods covered with soft hairs; flowers pink to purple.

7b. Flower heads spherical . . . . go to 8

8a (7b). Leaves with 3–10 pairs of pinnae . . . . go to 12

8b. Leaves with 1–3 (rarely 4) pairs of pinnae . . . . go to 9

9a (8b). Pods with prickles on margins and on flat surfaces; leaflets covered with silky hairs . . . .
Mimosa emoryana—shrub about 3 feet (1 m) tall; in Presidio and Brewster counties; 3–5 pairs of leaflets per pinna; leaflets to ⅛ inch (3 mm) long; flowers pink; pods eventually breaking into segments.

9b. Pods with prickles only on margins, or without prickles; leaflets not silky . . . . go to 10

10a (9b). 1–3 pairs of leaflets per pinna . . . .
Mimosa turneri—in Brewster, Presidio, and Hudspeth counties; leaves with 1 or 2 pairs of pinnae; flowers pink.

10b. Typically with 3 or more pairs of leaflets per pinna (but can have fewer) . . . . go to 11

11a (10b). Pod yellow or light to dark brown, often constricted between seeds, twisted and eventually breaking into segments . . . .
. . . . *Mimosa borealis* (6a.)

11b. Pods dark-colored, usually straight, not twisting and breaking into segments; leaves with 1–4 pairs of pinnae . . . .
. . . . *Mimosa texana* (6b.)

12a (8a). Leaflets 1/16 inch (1.5 mm) long; flowers white; pods with straight prickles along margin, and covered with soft hairs . . . .
Mimosa warnockii—In Jeff Davis and northern Presido counties; branches zigzagging; pods not breaking into segments.

12b. Leaflets 1/16–¼ inch (2–6 mm) long; flowers pink or white; pods with or without curved or straight prickles, and with few or no hairs . . .
. . . . *Mimosa aculeaticarpa* (*M. biuncifera*) (5a.)

254. Retama, Jerusalem Thorn, Paloverde, *Parkinsonia aculeata*
Fabaceae, Legume Family

Shrub or small tree to 33 feet (10.2 m) tall, with smooth green bark; armed with stout spines; weedy, escaping from cultivation, in fields and disturbed ground, along roadsides; common. LEAVES: Deciduous, alternate and clustered, twice pinnately compound, with 2 or 4 pinnae joined at their bases; whole leaf 4–20 inches (1–5 dm) long, appearing feathery; midrib flat; leaflets numerous, ⅛–¼ inch (4–7 mm) long; margins entire. FLOWERS: Spring–fall; bright yellow, often splotched with red; fragrant; about 1 inch (2.5 cm) wide; 5 petals. FRUIT: Papery pod 1–5½ inches (2.5–14 cm) long, constricted between hard seeds. RANGE: May be naturalized ornamental in southwestern U.S.; south to South America.

255. Paloverde, Retama, *Parkinsonia texana* (*Cercidium macrum, C. texanum*)
Fabaceae, Legume Family

Shrub to small tree about 6–15 feet (2–5 m) tall; bark smooth, green; branches slightly zigzagging and armed with solitary spines or spines not obvious; in dry brush, hills; common; with 2 varieties. LEAVES: Deciduous, alternate, twice pinnately compound, with 1 to 3 pairs of pinnae; 1 to 3 pairs of leaflets per pinna; leaflets ⅛–½ inch (3–12 mm) long; margins entire. FLOWERS: Spring–fall; bright yellow, 1 petal speckled with red; ¾ inch (2 cm) wide; 5 petals. FRUIT: Thin legume to 3 inches (8 cm) long, ⅛–½ inch (4–12 mm) wide. RANGE: Texas; Mexico.

256. Texas Ebony, Ebony Ape's Earring,
Ebano, *Ebenopsis ebano (Pithecellobium flexicaule)*
Fabaceae, Legume Family

Shrub or tree to 40 feet (12 m) tall, with paired spines; often with zigzag branching; in woods or thickets; common. LEAVES: Persistent in mild winters, alternate, twice pinnately compound, with 1–3 pairs of pinnae; 2–6 pairs of leaflets per pinna; leaflets oblong or oval, ¼–½ inch (7–12 mm) long; dark green; margins entire. FLOWERS: Spring–summer; pale yellow or creamy white, fragrant; in cylindric spikes ¾–2 inches (2–5 cm) long; protruding stamens give bottle brush effect. FRUIT: Thick woody pod 4–8 inches (1–2 dm) long, about 1 inch (2.5 cm) wide; seeds reddish, ½ inch (1.2 cm) long. RANGE: Texas; Mexico. NOTE: This popular landscape plant in South Texas is damaged in freezes. The hard wood is valued for carving.

(photos 257, 258) Honey Mesquite,
Prosopis glandulosa
Fabaceae, Legume Family

Shrub or small to medium tree to 52 feet (16 m) tall, usually with straight stout thorns (paired or single); widespread and abundant, particularly in disturbed grasslands; also in deserts, mountains, floodplains. LEAVES: Deciduous, alternate, twice pinnately compound, with 1 pair (sometimes 2 pairs) of pinnae; 5–24 pairs of leaflets per pinna; leaflets linear or oblong, ½–2½ inches (1.2–6.5 cm) long; margins entire. FLOWERS: Spring–summer; pale yellow; fragrant; protruding stamens form cylindrical spikes 1–5½ inches long (2.5–14 cm), resembling bottle brushes. FRUIT: Slender pod to 10 inches (2.5 dm) long; yellow, often with red splotches; pulp sweet, edible. RANGE: Southwestern U.S.; Mexico. NOTE: Though ranchers consider honey mesquite a nuisance because of its increased abundance, the plants provide valuable forage for livestock and wildlife as well as shelter for herbs and grasses. The wood is valued for carving and charcoal. RELATED SPECIES: 3 species with leaflets less than ½ inch (1.2 cm) long. *Prosopis laevigata*, in Nueces County, is rare. Two tornillos or screwbean mesquites, have pods that are 1–2 inches (1.2–5 cm) long and twisted into tight coils— *P. pubescens* (photo 258) in the Trans-Pecos, with cylindrical flower spikes, and *P. reptans* var. *cinerascens* in South Texas, with round yellow flower heads. The type of pods distinguishes *Prosopis* from *Acacia* (page 224) and *Mimosa* (page 230).

259. Black Locust, *Robinia pseudoacacia*
Fabaceae, Legume Family

Small to medium tree to about 45 feet (15 m) tall, with paired spines frequently at base of leaf and along branches; common, weedy. LEAVES: Deciduous, alternate, pinnately compound; leaflets numerous, generally elliptic, ½–2 inches (1.2–5 cm) long; margins entire. FLOWERS: Spring; white; fragrant; pea-type, in showy hanging clusters. FRUIT: Flat smooth pod 2–5 inches (5–12 cm) long. RANGE: Probably native to southeastern U.S.; grown throughout Texas, escaping cultivation in the eastern half. NOTE: All parts of the plant, including the beans, are poisonous. RELATED SPECIES: 2 shrubs—*Robinia neomexicana*, in the Guadalupe Mountains, has rosy to white flowers and hairy bean pods. In Northeast Texas, *R. hispida* has rose to purple flowers and densely hairy branches; probably introduced.

260. Rattlebush, Rattlepod, *Sesbania drummondii*
Fabaceae, Legume Family

Weak shrub or small tree about 2–9 feet (6–30 dm) tall; dying back in cold winters; often in sunny locations with sandy soil, along streams, in wetlands; locally abundant and weedy. LEAVES: Deciduous, alternate, pinnately compound; leaflets numerous, oblong or elliptic, ½–1¼ inches (1.2–3.5 cm) long; margins entire. FLOWERS: Summer–fall; yellow streaked with red; pea-type; in showy dangling clusters. FRUIT: 4-winged pod 2–3 inches (5–8 cm) long; seeds rattling inside dry pod. RANGE: Coastal states from Florida to Mexico. RELATED SPECIES: Brazil rattlepod (*Sesbania punicea*), with orange-red to purple flowers in spring, may escape cultivation in East Texas. NOTE: All parts of rattlepods are poisonous.

261. Eve's Necklace, *Sophora affinis*
Fabaceae, Legume Family

Shrub or small tree about 6–30 feet (2–10 m) tall; in dry, usually limestone soil, along fences, in disturbed ground, by streams, and on slopes; uncommon. LEAVES: Deciduous, alternate, pinnately compound; leaflets numerous, oval to elliptic, ¾–1½ inch (2–4 cm) long; margins entire. FLOWERS: Spring; white to rosy pink; fragrant; pea-type, in loose cascading clusters. FRUIT: Black leathery pod, pinched between seeds (resembling beads on necklace); seeds black. RANGE: Oklahoma to Louisiana. NOTE: The seeds may be poisonous.

262. Texas Mountain Laurel,
Mescalbean, Frijolillo, *Sophora secundiflora*
Fabaceae, Legume Family

Small shrub to medium-sized tree to 27 feet (8.3 m) tall; usually in limestone soil, in wooded hills, canyons, mountains; common. LEAVES: Evergreen, alternate, pinnately compound with 5–13 leaflets, leaflets oval, elliptic, or egg-shaped (widest near tip), 1–3 inches (2.5–7.5 cm) long; top dark green, shiny; margins entire. FLOWERS: Early spring; blue to purple (sometimes white); aroma similar to grape bubble-gum; pea-type, in showy cascading clusters. FRUIT: Plump woody pod 1–5 inches (2.5–12 cm) long, constricted between seeds; seeds hard, scarlet. RANGE: Texas, New

Mexico; Mexico. NOTE: A popular drought-tolerant landscape plant, Texas mountain laurel has poisonous beans, leaves, and flowers. RELATED SPECIES: *Sophora gypsophila*, a very rare small shrub in the Guadalupe Mountains, has smaller leaflets with silvery hairs; *S. tomentosa*, a small shrub along the southern coast, has yellow flowers and leaves velvety.

263. Chinquapin, *Castanea pumila*
Fagaceae, Beech Family

Shrub or medium-sized tree to about 50 feet (15 m) tall, with rough bark; in sandy woods, thickets; uncommon, with 2 varieties. LEAVES: Deciduous, simple, alternate; blade elliptic to lanceolate, 3–6 inches (7.5–15 cm) long; bottom white, usually woolly; veins straight, parallel; margins regularly toothed, teeth bristle-tipped. FLOWERS: Spring; catkins green, or showy and white to rusty-colored. FRUIT: Tiny edible nutlets enclosed in prickly hull ½–1½ inches (1.2–4 cm) in diameter. RANGE: Eastern U.S. RELATED SPECIES: Chestnut (*Castanea dentata*), an eastern tree not native to Texas, has been nearly eliminated by chestnut blight.

264. Beech, *Fagus grandifolia*
Fagaceae, Beech Family

Large-trunked tree to 132 feet (40.3 m), with smooth gray bark; in forests, usually near streams; common. LEAVES: Deciduous, simple, alternate; blade egg-shaped to elliptic, 3–6 inches (8–15 cm) long; veins straight and prominent; margins toothed. FLOWERS: Spring; inconspicuous; male flowers in round heads, females in small clusters on same tree. FRUIT: Triangular thin-shelled edible nuts ½ inch (1.2 cm) long, enclosed in prickly hull. RANGE: Eastern U.S. to Nova Scotia.

(photos 265 through 279) Oak, *Quercus* **species (see Key)**
Fagaceae, Beech Family

Shrubs or trees; leaf buds usually clustered at twig tips. LEAVES: Deciduous (dead leaves often hang on tree through winter), nearly evergreen (leaves green through winter, falling off in early spring), or truly evergreen; simple, alternate. FLOWERS: Most species flower in early spring; inconspicuous catkins; pollen allergenic. FRUIT: Acorn (1-seeded nut capped with a woody cup). RELATED SPECIES: Of 39 species in Texas, 29 of the most common are included here; 6 rare and endangered species are not included. NOTE: Some species produce sweet acorns; the tannic acid can be leached out with boiling water, and the nutmeats can be dried and ground into flour. Various parts used as dyes and mordants.

KEY TO OAKS

Oaks hybridize freely, causing much difficulty in identification. This key is designed for non-hybrid specimens.

1a. East of the Pecos River, or may overlap onto the edge of the Trans-Pecos (but not in mountains) . . . . go to 8

1b. In mountains, foothills, and canyons of the Trans-Pecos . . . . go to 2

2a (1b). Leaf margin lobed, lobes extending about halfway to midrib or deeper; blade 2–7 inches (5–17.5 cm) long . . . . go to 4

2b. Not as above . . . . go to 3

3a (2b). Leaves small, blade generally less than 3½ inches (cm) long; some species nearly evergreen, with leaves thick and leathery . . . .
. . . . go to 5 (several rare Trans-Pecos oaks not included)

3b. Leaves broad, larger, 3–10 inches (8–25 cm) long, deciduous, thin; leaves usually have 8–12 veins on each side of midrib, each vein ending in a round or pointed tooth or shallow lobe; acorns fairly large, to 1 inch (2.5 cm) long; in woods, dry limestone soil, in scattered populations in Trans-Pecos, Central, and Northeast Texas . . . .
. . . . chinkapin oak, *Quercus muehlenbergii* (photo 273)

4a (2a). Blade deeply lobed, lobes rounded, without any bristly hairs on lobe tips; thicket-forming low shrub or fairly large tree at high elevations
. . . . Gambel oak, *Quercus gambelii*
(hybridizes with *Q. pungens* and various other species).

4b. Leaf blade with shallow to deep lobes, lobes pointed and tipped with bristly hairs; leaf top shiny; petiole to 1 inch (2.5 cm) long; young bark smooth and gray, older bark dark and rough; leaves deciduous, turning red or gold in fall, to nearly evergreen; trees common to abundant in mountains, from the Davis Mountains south, with an isolated population in Val Verde County at lower elevations; acorns small, to ⅗ inch (1.5 cm) long ....
.... Chisos red oak, *Quercus gravesii* (photo 268)
(hybridizes with *Q. emoryi*)

5a (3a). Leaf bright green, shiny, narrowly lanceolate to ovate or oblong; 1–4 inches (2.5–10 cm), but typically less than 2 inches (5 cm) long; margins with spine-tipped teeth (sometimes entire or with shallow lobes), usually with a tuft of hairs on leaf bottom at base of midrib; rest of leaf hairless to somewhat hairy, but hairs not dense; petiole less than ½ inch (1.2 cm) long; leaves nearly evergreen, turning yellow in late winter; small to large trees, common to abundant in mountains and canyons, from Davis Mountains south .... Emory oak, *Quercus emoryi* (photo 266)
(hybridizes with *Q. gravesii* and others)

5b. Not as above ....go to 6

6a (5b). Leaf top shiny, green; bottom white or gray, and densely hairy, woolly, or velvety; leaves oblong or elliptic; margins entire or with a few sharp-pointed teeth; persistent to evergreen; petiole to ¼ inch (6 mm) long; acorns small, to ⅗ inch (1.5 cm) long; low shrub or small tree, widespread but scattered in western half of Texas, mainly in limestone .... scrub oak, shin oak, *Quercus mohriana*
(hybridizes with *Q. grisea*)

6b. Not as above ....go to 7

7a (6b). Large shrub to large tree; leaf grayish green, bloom makes leaf appear dull; blade ovate, elliptic, or oblong, usually less than 2 inches (5 cm) long; margins entire or with a few sharp-pointed teeth but not lobed; hairs minute, rarely dense (hand lens); petiole to ⅜ inch (1 cm) long; persistent to nearly evergreen; common to abundant in mountains, mainly in igneous areas, but also in limestone .... gray oak, *Quercus grisea* (photo 269)
(hybridizes with *Q. mohriana, Q. pungens*, and others)

7b. Usually a low thicket-forming shrub, occasionally a medium-sized tree; gray bark shaggy; leaf top shiny, green to gray-green, smooth to sandpapery rough; leaf surfaces with minute hairs (hand lens), hairs sparse or dense but bottom not white and woolly (though may be lighter color than leaf top); margins with numerous spine-tipped teeth or shallow lobes (sometimes entire), surface flat or margins undulating or rolled under; petiole to ⅜ inch (1 cm) long; persistent to nearly evergreen; acorns small, ½–¾ inch (1.2–2 cm) long; abundant in the Trans-Pecos; extends to the Edwards Plateau .... shin oak, scrub oak *Quercus pungens* (hybridizes with several other species)

8a (1a). In East Texas, to Austin or Dallas, or in South Texas, along the coast .... go to 19

8b. From Austin and Dallas west, in Central or Northwest Texas, to the edge of the Trans-Pecos .... go to 9

9a (8b). Leaf triangular, with broad tip 3–7 inches (7.5–17.5 cm) wide, margins entire or 3-lobed in upper half; lobes tipped with bristly hairs (may break off with age); leaves deciduous, 2–7 inches (5–18 cm) long (West Texas leaves smaller than in wetter East Texas); acorns small, to 1 inch (2.5 cm) long; tree to 51 feet (15.7 m) tall; in sand, gravel, or clay, from East Texas to the edge of the Panhandle .... blackjack oak, *Quercus marilandica* (photo 272) (hybridizes with *Q. buckleyi, Q. nigra,* and others)

9b. Not as above .... go to 10

10a (9b). Thicket-forming shrub often less than 3 feet (1 m) tall, occasionally a small tree; in deep sand; in the Panhandle, through the south Panhandle Plains and western Edwards Plateau to Loving County; acorns fairly large, cups ½–1 inch (1.2–2.5 cm) wide; leaves deciduous, thick; shape variable, generally oblong to ovate; ¾–4 inches (2–10 cm) long, top shiny, bottom often densely hairy (hairs gray or yellow); leaf margins entire, wavy, toothed, or deeply lobed .... Havard shin oak. *Quercus havardii* (hybridizes with *Q. mohriana, Q. stellata*)

10b. Not as above .... go to 11

11a (10b). Margins of most leaves deeply pinnately lobed, lobes extending at least halfway to midrib .... go to 17

11b. Margins entire, toothed, or lobed; lobes shallow (rarely a few leaves with lobes cut more than halfway to midrib) .... go to 12

12a (11b). Leaves broad, 3–10 inches (8–25 cm) long; usually 8–12 veins on each side of midrib, each vein ending in a round or pointed tooth or shallow lobe; deciduous; acorns fairly large, to 1 inch (2.5 cm) long; in woods, dry limestone soil, in scattered populations in the Trans-Pecos, Central, and Northeast Texas .... chinkapin oak, *Quercus muehlenbergii* (photo 273)

12b. Leaves generally small, less than 5 inches (1.3 dm) long, not as above .... go to 13

13a (12b). Leaf top usually green, shiny; bottom whitish or gray, densely hairy, woolly or velvety; low shrub or small tree, mainly in limestone, widespread but scattered in western half of Texas; leaves to 3 inches (8 cm) long, persistent to evergreen; margins entire or with a few sharp-pointed teeth . . . . scrub oak, shin oak, *Quercus mohriana*

13b. Bottom of mature leaf not covered with woolly or velvety hairs . . . . go to 14

14a (13b). Margins usually entire (occasionally with a few sharp-pointed teeth).

> Plateau live oak, *Quercus fusiformis*—Shrub to large tree, common on limestone outcrops or granite hills in Central, North Central, and South Texas; bark hard, dark; blade oblong, elliptic, lanceolate, or oblanceolate, 1–6 inches (2.5–15 cm) long, thick, nearly evergreen; leaf top usually dark green and shiny, leaf bottom pale; acorns somewhat fusiform, ½–1 inch (1.2–2.5 cm) long. SIMILAR SPECIES: intergrades with *Q. virginiana* (see 28a) in deeper soils from edge of the Edwards Plateau to the coast.

14b. Margins wavy or with shallow rounded lobes or with sharp-pointed teeth or lobes (occasionally entire) . . . . go to 15

15a (14b). Small to medium tree on limestone in Central Texas and in isolated populations in the Trans-Pecos; leaf top with a bluish or grayish cast, usually glaucous; leaf margins wavy or with shallow rounded lobes (occasionally entire or rarely deeply lobed or with a few spine-tipped teeth); deciduous; bark gray, may be loose, but not as shaggy as on species below . . . . Lacey oak, *Quercus laceyi* (photo 270) (*Q. glaucoides*)

15b. Thicket-forming low shrubs or small trees in Central Texas; bark gray, loose, shaggy . . . . go to 16

16a (15b). Leaf margins wavy or with shallow rounded lobes (occasionally entire); leaves oblong to fairly broad, usually widest above the middle . . . .

> Shin oak, *Quercus sinuata* var. *breviloba* (photo 277)—Common on limestone in Central and North Central Texas; leaves deciduous, 1–4 inches (2.5–10 cm) long; leaf top gray or green, often shiny, bottom paler; acorns ½–1 inch (1.2–2.5 cm) long. SIMILAR SPECIES: *Quercus sinuata* (see 25a), in East and Central Texas, is a solitary, often large tree.

16b. Teeth and shallow lobes sharp-pointed (margins sometimes entire); leaves generally narrow, oblong to elliptic (sometimes broader) . . . .

Vasey oak, *Quercus vaseyana* (photo 276)—Common on rocky soil in Trans-Pecos and Central Texas; leaves evergreen or persistent, usually less than 3 inches (7.5 cm) long.

17a (11a). Lobes and teeth with bristly hairs on tips . . . .

Texas oak, Spanish oak, *Quercus buckleyi* (*Q. texana*) (photo 265)—Small to large tree, common in dry limestone ridges or stream bottoms, or sand or gravel, endemic to the Edwards Plateau and North Central Texas; bark smooth and gray, to rough and dark; leaves deciduous, thin, turning red and yellow in fall, 2–5½ inches (5–14 cm) long; 2–4 lobes on each side of leaf, lobes extending ½–⁹⁄₁₀ the distance to midrib; acorns to ¾ inch (2 cm) long, bottom rounded; cup fairly deeply rounded; hybridizes with *Q. marilandica*. SIMILAR SPECIES: Along the edge of the Edwards Plateau it intergrades with *Quercus shumardii*, which generally has broader and shallower lobes (see 31a).

17b. Lobes rounded, no bristles on tips . . . . go to 18

18a (17b). Leaves usually very large, 4–14 inches (1–3.5 dm) long; acorns large, to 2 inches (5 cm) long; cup to 2½ inches (6.5 cm) wide, with fringed edge; large trees in East, Central, and North Central Texas . . . . bur oak, *Quercus macrocarpa* (photo 271)

18b. Leaf smaller, 2–6 inches (5–15 cm) long; acorns small, ½–1¼ inches (1.2–3 cm) long; acorn cup to 1 inch (2.5 cm) wide; middle lobes of leaves often largest, may give the leaf a cross-shaped appearance . . . .

Post oak, *Quercus stellata* (photo 278)—Small to large tree in dry upland woods, in sand, gravel, or clay; common and widespread from East and Central Texas to the edge of the Panhandle; bottom of mature leaf sparsely hairy to nearly hairless. SIMILAR SPECIES: *Quercus drummondii* and *Q. margaretta* may be varieties of *Q. stellata* (see 33a).

19a (8a). Leaves triangular, with broad tip; margins entire or 3-lobed in upper half, lobes tipped with bristly hairs unless broken off (water oak leaves are highly variable, but usually some leaves are as above) . . . . go to 20

19b. Leaf not as above . . . . go to 22

20a (19a). Leaves persistent to nearly evergreen; blade narrow, tips generally less than 2½ inches (6.5 cm) wide; twigs with few or no hairs; petioles to ¼ inch (7 mm) long . . . .

Water oak, *Quercus nigra* (photo 274)—Tree to 146 feet (45 m) tall, East Texas to Austin; bark hard, dark, smooth or furrowed; leaves 2–7 inches (5–18 cm) long; blade variable in shape, can be triangular, elliptic, oblong, or diamond-shaped; leaf margins entire, 3-lobed, or sometimes pinnately lobed; leaf bottom may have tufts of hairs in axils of veins; hybridizes with *Q. laurifolia*, *Q. marilandica*, *Q. phellos*, and others.

20b. Leaves deciduous; blade broad, tips 2½–7 inches (6.5–18 cm) wide; twig and petiole often hairy .... go to 21

21a (20b). Leaf surfaces usually nearly hairless when older (may be densely hairy when young), bottom with tufts of hairs in axils of veins; petioles less than ⅝ inch (1.5 cm) long; leaves 2–7 inches (5–18 cm) long; margins entire or 3-lobed near tip ....
.... blackjack oak, *Quercus marilandica* (see 9a; photo 272)

21b. Leaf bottom usually velvety, even when mature; hairs yellow or brown; margins quite variable, 3-lobed or deeply pinnately lobed (not entire); petioles ⅜–2 inches (1–5 cm) long; leaves 3–12 inches (8–30 cm) long ....
.... southern red oak, *Quercus falcata* (see 30a; photo 267)

22a (19b). On most leaves, margins deeply lobed, lobes extending at least halfway to midrib .... go to 29

22b. On most leaves, margins not deeply lobed .... go to 23

23a (22b). Margins of most leaves entire or wavy or with a few shallow rounded lobes (occasionally with a few irregularly spaced spine-tipped teeth); leaves generally 1–6 inches (2.5–15 cm) long
.... go to 24

23b. Leaf with about 12–15 veins on each side or midrib, each ending in a rounded or sharp-pointed tooth or lobe; blade usually large and broad, 4–10 inches (1–2.5 dm) long ....
Chestnut oak, (*Quercus michauxii*)—Large tree in moist woods, bottomlands east of Houston; bark soft gray; leaf bottom usually velvety; acorns large, cup to 1½ inches (4 cm) wide. SIMILAR SPECIES: Chinkapin oak (*Q. muehlenbergii*, photo 273), in dry limestone soils in scattered populations in West, Central, and Northeast Texas, has leaf bottom smooth to slightly fuzzy and usually 8–12 veins on each side of leaf. *Quercus alba* (see 35b) sometimes has shallow lobes, but leaf bottoms are not velvety.

24a (23a). Margins of most leaves entire (with some variability in each species) .... go to 26

24b. Margins usually wavy or with rounded lobes (occasionally entire or a few teeth spine-tipped) .... go to 25

25a (24b). Leaves generally widest above the middle; lobes not tipped with bristly hairs ....
Bastard oak, *Quercus sinuata*—Large tree, uncommon in East Texas to the edge of the Edwards Plateau, along streams; bark gray, loose, shaggy; leaves deciduous, 2–6 inches (5–15

cm) long, top shiny, dark green when mature; acorns small, to 1¼ inches (3 cm) long.

25b. Leaves mostly entire, elliptic or diamond-shaped or widest near base (sometimes widest above the middle, margins wavy or lobed); lobes, when present, tipped with bristly hairs (may break off with age) . . . . laurel oak, *Quercus laurifolia* (see 28b)

26a (24a). Leaves narrow, more than 3 times longer than wide . . . .
Willow oak, *Quercus phellos* (photo 275)—Medium to large tree in moist woods in East Texas; bark hard, somewhat smooth; leaves deciduous, elliptic to narrowly lanceolate or oblanceolate, 2–5 inches (5–12.5 cm) long, ⅜₆–1½ inches (5–40 mm) wide; top dark green and often shiny, bottom velvety to nearly hairless; margins entire or with a few teeth tipped with bristly hairs (young leaves rarely with a few shallow lobes); acorns nearly round, to ⅝ inch (1.5 cm) long; hybrids with *Q. nigra* or *Q. falcata* may be deeply lobed.

26b. Leaves about 2 or 3 times longer than wide . . . . go to 27

27a (26b). Leaf bottom velvety or woolly, gray (rarely sparsely hairy) . . . . .
Sandjack, bluejack, *Quercus incana*—Shrub or small tree in dry sandy woods in East Texas to edge of Edwards Plateau; leaves deciduous, elliptic, oblong, or lanceolate, 2–5 inches (5–12.5 cm) long; leaf top shiny; margins entire or with a few teeth tipped with bristly hairs; acorns nearly round, to ⅝ inch (1.5 cm) long; hybridizes with *Q. marilandica, Q. nigra.*

27b. Bottom not velvety or woolly . . . . go to 28

28a (27b). Leaf tip and teeth may be sharp-pointed but not with distinct bristly hairs; leaf top dark green, shiny; leaf bottom pale, covered with minute hairs (hand lens) . . . .
Coastal live oak, *Quercus virginiana* (photo 279)—Small to large, widely branching tree, in woods along the coast and in South Texas; bark dark and deeply furrowed; leaves nearly evergreen; obovate, oblong, elliptic, lanceolate, or oblanceolate, 1–6 inches (2.5–15 cm) long; margins entire, occasionally with a few sharp-pointed teeth; acorns ovate to fusiform, to 1 inch (2.5 cm) long. SIMILAR SPECIES: Intergrades with *Q. fusiformis* (see 14a) from the coast to edge of Edwards Plateau.

28b. Leaf tipped with bristly hairs (may break off with age); leaf top shiny, surfaces hairless to sparsely hairy, or with hairs along base of midrib . . . .
Laurel oak, *Quercus laurifolia*—Large tree in wet woods in Southeast Texas; bark gray or black, hard; leaves deciduous to nearly evergreen, 2–6 inches (5–15 cm) long, blade gener-

ally elliptic or diamond-shaped but highly variable, may be wide near base or tip; margins generally entire, but may be lobed or have several teeth near tip; acorns ovate to nearly round, to ⅝ inch (1.5 cm) long; hybridizes with *Q. falcata, Q. nigra, Q. phellos*.

29a (22a). Lobes rounded or pointed, with no bristly hairs; leaves thin or thick, deciduous . . . . go to 32

29b. Lobes and teeth with pointed tips ending in bristly hairs; leaves thin, deciduous . . . . go to 30

30a (29b). Leaf bottom usually velvety or fuzzy, hairs often yellowish or brownish; top shiny, velvety to hairless; leaf base often U- or V-shaped; lobes highly variable . . . .
Southern red oak, *Quercus falcata* (photo 267)—Large tree, common in moist woods in East Texas; leaves 3–12 inches (8–30 cm) long, petiole ⅜–2 inches (1–5 cm) long; 3 leaf varieties—(1) with a 3-lobed broadened tip, (2) deeply pinnately lobed, with numerous lobes, (3) with few deep side lobes and an elongated terminal lobe; leaves turning orange to reddish brown in fall; acorns to ⅝ inch (1.5 cm) long, bottom round or flattened; hybridizes with *Q. nigra, Q. phellos*, and others.

30b. Leaf bottom not velvety, may be densely or sparsely hairy or with tufts of hairs only in leaf axils (hand lens); top shiny, hairless to sparsely hairy; leaf base may be rounded but not U-shaped; leaf pinnately lobed, with 2–4 lobes on each side of midrib . . . go to 31

31a (30b). Buds hairless or with few hairs; leaf never densely hairy; bottom sparsely hairy or with tufts of hairs restricted to axils of veins; acorns to 1 inch (2.5 cm) long, bottom flattened; acorn cup shallow, flattened or only slightly rounded; cup scales not loose or protruding . . . .
Shumard oak, *Quercus shumardii*—Large tree from East Texas to edge of Edwards Plateau, in moist woods or stream bottoms; leaves 3–8 inches (8–20 cm) long, turning red or yellow in fall; lobes extending ½ to ¾ the distance to midrib. SIMILAR SPECIES: Intergrades with Texas oak (*Q. buckleyi*, see 17a). *Quercus texana*, less common in East Texas, has lobes cut deeper than in *Q. shumardii*, rounded deep acorn cups, and leaves that are not colorful in the fall.

31b. Buds densely hairy; leaf bottom densely hairy (hairs brownish) or with hairs restricted to midrib and leaf axils; acorns elliptic to nearly round, to ¾ inch (2 cm) long; acorn cups rounded, fairly deep, yellow inside; tips of cup scales loose or protruding . . . . . .

Black oak, *Quercus velutina*—Medium to large tree in upland woods in East Texas; uncommon; leaves to 10 inches (2.5 dm) long; inner bark may be bright yellow to orange, used for dying wool.

32a (29a). Leaves often large, 3–14 inches (8–35 cm) long; lobes rounded or pointed . . . . go to 34

32b. Leaves smaller, 2–6 inches (5–15 cm) long; lobes rounded; middle lobes often the largest, may give the leaf a cross-shaped appearance (all the post oaks may be varieties of *Q. stellata*) . . . go to 33

33a (32b). Low shrub or small tree, in deep sand in scattered populations in East, North Central, and Central Texas; leaves thin, bottom velvety to nearly hairless . . . .
Sand post oak, *Quercus margaretta*. SIMILAR SPECIES: Drummond post oak (*Q. drummondii*), a small to medium-sized tree in sand in East Texas; has thick, leathery leaves.

33b. Small to large trees; bottom of mature leaves sparsely hairy to nearly hairless . . . .
Post oak, *Quercus stellata* (photo 278)—In dry upland woods, in sand, gravel, or clay, common and widespread from East and Central Texas to the edge of the Panhandle; leaves deciduous but thick; acorn small, ½–1¼ inches (1.2–3 cm) long; var. *similis*, a tree in wet bottomlands in East Texas, has thin leaves that are not usually cross-shaped.

34a (32a). Found mainly west of Houston, in East, North Central, and Central Texas, to parts of South Texas; mainly in limestone soils; acorns large, to 2 inches (5 cm) long; cup 1–2½ inches (2.5–6.5 cm) wide, fringed with coarse hairs . . . .
Bur oak, *Quercus macrocarpa* (photo 271)—Large tree in woods, near streams; leaves 4–14 inches (1–3.5 dm) long; lobes rounded; leaf top nearly hairless; bottom densely hairy and gray or sparsely hairy and green.

34b. Found mainly east of Houston; in forests; acorns smaller, to 1 inch (2.5 cm) long; cup to 1¼ inches (3 cm) wide, not fringed . . . . go to 35

35a (34b). Medium to large tree in swamps and wetlands; acorns usually hemispheric, the nut generally fully enclosed by the cup; cup to ¾ inch (2 cm) deep; immature acorns of bur oak (34a) or white oak (35b) may look similar; leaf lobes usually pointed; leaves to 10 inches (2.5 dm) long . . . . overcup oak, *Quercus lyrata*

35b. Small to large tree in well-drained soils; acorn cup fairly shallow, enclosing less than half of nut; leaf lobes rounded, shallow or deep; leaves to 9 inches (2.2 dm) long, deciduous, turning reddish in fall . . . . white oak, *Quercus alba*

280. Ocotillo, *Fouquieria splendens*
Fouquieriaceae, Ocotillo Family

Bizarre desert shrub about 3–30 feet (1–9 m) tall; vertical branches spiny, wandlike, emerging from root crown; branches ¼–2½ inches (6–60 mm) in diameter; on dry rocky hills, in deserts; locally abundant. LEAVES: Usually absent except after rains; simple, alternate or clustered; blade elliptic to obovate, ½–2 inches (1.2–5 cm) long; margins entire. FLOWERS: Spring; bright red; tubular, to ¾ inch (2 cm) long; in flamboyant clusters at branch tips; red stamens protruding. FRUIT: 3-celled woody capsule; many tiny seeds. RANGE: Texas to California; Mexico. NOTE: Flowers are edible.

281. Silktassel, *Garrya ovata (G. lindheimeri)*
Garryaceae, Garrya Family

Small shrub to about 15 feet (4.5 m) tall; shade-tolerant; on rocky hills, mountains, canyons, and along streams; common. LEAVES: Evergreen, simple, opposite; blade elliptic, oval, or egg-shaped, 1–3 inches (2.5–8 cm) long, leathery; bottom often densely hairy (hand lens); top may become shiny, smooth with age; margins entire, flat or slightly undulating (subsp. *lindheimeri*) or strongly undulating (subsp. *goldmanii*). FLOWERS: Early spring; in catkins; dioecious. FRUIT: Purple drupe, usually dry, to ⅜ inch (1 cm) in diameter. RANGE: Texas, New Mexico; Mexico. RELATED SPECIES: *Garrya wrightii*, in the mountains and canyons of the Trans-Pecos, has leaves that are not undulating and usually not densely hairy.

282. Witch Hazel, *Hamamelis virginiana*
Hamamelidaceae, Witch Hazel Family

Shrub or small tree to 26 feet (8 m) tall; in partial shade; in woods; uncommon. LEAVES: Deciduous, simple, alternate; blades oval to nearly round, 2–6 inches (5–15 cm) long, with prominent straight veins, uneven base; margins wavy or with rounded teeth. FLOWERS: Fall; yellow; linear petals to ¾ inch (2 cm) long; in tassel-like clusters, often covering tree after leaves drop. FRUIT: Woody capsule ½ inch (1.2 cm) long; seeds shiny, black. RANGE: Eastern U.S.; Canada. RELATED SPECIES: *Hamamelis vernalis*, in East Texas, blooms in late winter and early spring. NOTE: A distilled extract of the twigs yields a multi-purpose home remedy.

283. Sweetgum, *Liquidambar styraciflua*
Hamamelidaceae, Witch Hazel Family

Large tree to 108 feet (33 m) tall, often with corky wings on branches; in moist sandy or gravelly soil, in forests or fields; common. LEAVES: Deciduous, simple, alternate; palmately lobed, star-shaped with 3–7 lobes; blade 3–7 inches (8–18 cm) long and wide; turning red in fall; margins toothed. FLOWERS: Spring; green round heads ¼ inch (6 mm) in diameter. FRUIT: Spiked, woody ball 1½ inch (4 cm) in diameter, on long stem. RANGE: Eastern U.S.; Central America. NOTE: This fast-growing landscape tree has several horticultural varieties. The sap is used in pharmaceuticals and perfumes.

284. Red Buckeye, *Aesculus pavia* **var.** *pavia*
Hippocastanaceae, Horse Chestnut Family

Shrub or small tree to about 30 feet (9 m) tall; in sandy or rocky soil, in woods and hills, along streams. LEAVES: Falling by mid-summer; opposite, palmately compound; with 5 leaflets (rarely 7); leaflets oblanceolate to elliptic, 3–7 inches (7.5–17.5 cm) long; margins sharp-toothed. FLOWERS: Early spring; red (or yellow-streaked); about 1 inch (2.5 cm) long, 5-lobed; in large erect panicles. FRUIT: Tan leathery capsule; seeds about 1 inch (2.5 cm) in diameter, brown, with a light scar. RANGE: Southeastern U.S. RELATED SPECIES: *Aesculus pavia* var. *flavescens*, on the Edwards

Plateau, has yellow flowers. Two varieties of *A. glabra* have pale yellow or greenish flowers and usually spiny capsules: Ohio buckeye (*A. glabra* var. *glabra*), with 5–7 leaflets, in Northeast Texas; and Texas buckeye, (*A. glabra* var. *arguta*), with 7–11 leaflets, in Central, East and Northeast Texas. Hybrids result in various flower color patterns. NOTE: All parts of these plants are deadly poisonous.

285. St. Andrew's Cross, *Hypericum hypericoides*
   (Ascyrum hypericoides) Clusiaceae
   (Hypericaceae), St. John's Wort Family

Shrub to 4 feet (1.2 m) tall; in sandy soils, forests, grasslands, bogs; uncommon. LEAVES: Evergreen, simple, opposite; blade variable, narrowly linear to broadly oblanceolate or egg-shaped, to 1¼ inch (3 cm) long, ⅛–¼ inch (4–8 mm) wide; margins entire. FLOWERS: Late spring–fall; yellow; about ⅜ inch (1 cm) wide; 4 petals. FRUIT: Small capsule. RANGE: Eastern U.S.; Central America. RELATED SPECIES: St. Peter's Wort (*H. crux-andrae*), with larger leaves and flowers, is less widespread in East Texas.

286. Pecan, Nogal Pecanero,
   Nuez encarcelada, *Carya illinoinensis*
   Juglandaceae, Walnut Family

Large tree to 90 feet (27.5 m) tall with shaggy bark; in moist, rich woods, bottomlands and along waterways; common. LEAVES: Deciduous, alternate, pinnately compound with 9–17 leaflets; leaflets elliptic or lanceolate, often curving, 3–8 inches (8–20 cm) long; aromatic; margins toothed. FLOWERS: Spring; catkins; pollen allergenic. FRUIT: Edible nut, ¾–3½ inches (2–9 cm) long; shell smooth, oblong to oval; hull splitting open when mature. RANGE: South-central U.S.; Mexico. NOTE: Pecan is the state tree of Texas. Many horticultural varieties exist with various sized nuts. Leaves and woody parts make excellent dyes.

287. Black Hickory, Nogal, *Carya texana*
Juglandaceae, Walnut Family

Small to large tree to 135 feet (41.6 m) tall; bark dark, deeply furrowed; buds yellowish; young twigs and leafstalks may be covered with reddish or yellowish hairs, usually hairless when older; on rocky slopes, in sandy soil, in woods; common. LEAVES: Deciduous, alternate, pinnately compound, with 5–7 (rarely 9) leaflets; leaflets lanceolate, egg-shaped, or elliptic, 2–6 inches (5–15 cm) long; aromatic; margins sharp-toothed. FLOW-ERS: Spring; in catkins. FRUIT: Edible nut in egg-shaped to nearly round shell; hull 1–2 inches (2.5–5 cm) long, splitting open when mature. RANGE: Central U.S. RELATED SPECIES: This is the most widespread of 8 Texas hickories. Leaves and woody parts make excellent dyes.

288. Little Walnut, Nogal, *Juglans microcarpa*
Juglandaceae, Walnut Family

Shrub or small tree to 50 feet (15.4 m); along streams, in canyons and mountains; common. LEAVES: Deciduous, alternate, pinnately compound, with 15 or more leaflets; leaflets lanceolate, 2–5 inches (5–13 cm) long, ¼–1 inch (6–25 mm) wide; margins finely toothed to nearly entire. FLOWERS: Spring; in catkins. FRUIT: Edible nut in woody shell; husk round, ½–1½ inches (1.2–4 cm) in diameter, not splitting open. RANGE: Oklahoma to New Mexico; Mexico. RELATED SPECIES: Arizona walnut (*J. major*), a medium-sized tree in Central, North-Central and Trans-Pecos Texas, usually has fewer than 15 leaflets; the husks are usually 1–1½ inches (2.5–4 cm) in diameter. Eastern black walnut (*Juglans nigra*) is a large tree in the eastern half of Texas, with husks 1½–2½ inches (4–6.5 mm). The chambered pith distinguishes walnuts from other trees with similar leaves. NOTE: leaves and woody parts make an excellent dye.

289. Allthorn, Junco, *Koeberlinia spinosa*
  Capparaceae, Caper Family
  (Koeberliniaceae, Allthorn Family)

Many-branched shrub to 11 feet (3.4 m) tall; spine-tipped secondary branches about 1–4 inches (2.5–10 cm) long, smooth, green, alternating (one variety with slender spine-tipped branchlets often less than 1 inch , or 2.5 cm, long); in arid soil, in deserts and mountains; uncommon. LEAVES: usually absent; scalelike, minute. FLOWERS: Spring–fall; whitish or creamy yellow; inconspicuous. FRUIT: Brown to black fleshy berries, ¼ inch (6 mm) in diameter, nearly round. RANGE: Texas to Arizona; Mexico. SIMILAR SPECIES: Crucifixion thorn (*Holacantha stewartii*), in Brewster and Presido counties, is less than 3 feet (1 m) tall and has red, tan, or green fruit that is flattened and egg-shaped; junco (*Adolphia infesta*), in the Trans-Pecos, differs by having paired branches. The leaves of both species, when present, are tiny and linear to oblong.

290. Range Ratany, *Krameria erecta*
  (K. parvifolia, K. glandulosa)
  Krameriaceae, Ratany Family

Small shrub typically less than 3 feet (1 m) tall; twigs and leaves covered with white or gray hairs; twigs somewhat spine-tipped; in arid brush, desert hills, mountains; common. LEAVES: Deciduous, simple, alternate (paired just below flowers); blade linear or oblong, ⅙–½ inch (4–12 mm) long; margins entire. FLOWERS: Spring–fall; reddish purple; ⅜–¾ inch (1–2 cm) in diameter; 4 or 5 sepals, colorful, egg-shaped; petals smaller, inconspicuous, unequally shaped. FRUIT: Round pod, ¼ inch (7 mm) wide, covered with tiny spines. RANGE: Texas to California; Mexico. RELATED SPECIES: *Krameria grayi*, whose range overlaps, has leaves to 1 inch (2.5 cm) long and colorful linear or narrowly lanceolate sepals, bending downward; *K. ramosissima*, with leaves ¹⁄₁₆–¼ inch (2–7 mm) long, grows near the Rio Grande from Brewster to Hidalgo counties. See *Krameria lanceolata*, page 139.

291. Shrubby Blue Sage, Mejorana, *Salvia ballotiflora*
Lamiaceae, Mint Family

Small aromatic shrub to about 6 feet (2 m) tall, with slender square stems; in dry brush, rocky canyons; uncommon. LEAVES: Deciduous, simple, opposite, blade triangular to egg-shaped or elliptic, ¼–1½ inches (6–40 mm) long; bottom white, hairy; margins toothed. FLOWERS: Nearly year-round; blue or purple; about ½ inch (1.2 cm) long, bilateral; sepals hairy and dotted with minute glands (hand lens). RANGE: Texas; Mexico. NOTE: The leaves can be used as a spice. RELATED SPECIES: See *Salvia* species, pages 43–44.

292. Red Bay, *Persea borbonia*
Lauraceae, Laurel Family

Aromatic shrub or slender tree 10–52 feet (3–16 m) tall, often thicket-forming; in woods and swamps, by streams and shores, usually in sand; common. LEAVES: Evergreen, simple, alternate; blade elliptic or lanceolate, 2–8 inches (5–20 cm) long; shiny above, leathery, aromatic; margins entire. FLOWERS: Spring–summer; pale yellow; tiny, inconspicuous, bell-shaped. FRUIT: Blue-black drupe about ½ inch (1.2 cm) long. RANGE: Eastern U.S. We have seen an isolated population in Travis County. NOTE: The dried leaves can be used to season soups and sauces.

293. Sassafras, *Sassafras albidum*
Lauraceae, Laurel Family

Small to large tree to 62 feet (19 m) tall, with lemon scented leaves and aromatic bark; in sandy woods, disturbed areas; common. LEAVES: Deciduous, simple, alternate; blade elliptic to egg-shaped,

3–7 inches (7–18 cm) long; different leaf types occur on one tree—margin entire and unlobed, or margin lobed on one or both sides; turning sunset colors in fall. FLOWERS: Spring; greenish yellow; tiny; dioecious. FRUIT: Fleshy dark blue drupe ¼–½ inch (6–12 mm) long. RANGE: Eastern U.S. NOTE: The leaves are used as a spice, the root bark for tea. The oil from the bark is reportedly carcinogenic.

294. Woolly Butterfly Bush, *Buddleja marrubiifolia*
  Buddlejaceae, Butterfly-bush Family
  (Loganiaceae, Logania family)

Shrub to about 6 feet (2 m) tall; usually in dry limestone hills, canyons; uncommon. LEAVES: Evergreen, simple, opposite; blade oval or egg-shaped to nearly round, ¼–2 inches (0.7–5 cm) long; surfaces densely hairy, feltlike, hairs gray, yellowish, or reddish; margins round-toothed, scalloped. FLOWERS: Spring–summer; yellow, red, or orange; minute; in tight rounded heads ½ inch (1.2 cm) in diameter. FRUIT: Rounded pod with numerous tiny seeds. RANGE: Texas; Mexico. RELATED SPECIES: The flower color distinguishes this from 4 other Texas species. *B. racemosa*, with creamy flowers, is endemic to Edwards Plateau.

295. Southern Magnolia, *Magnolia grandiflora*
  Magnoliaceae, Magnolia Family

Large tree to 112 feet (34.5 m) tall; in moist woods and along streams; common. LEAVES: Evergreen, simple, alternate; blade stiff, broad, elliptic, 3–10 inches (8–25 cm) long; shiny, dark green above; rusty hairs below; margins entire. FLOWERS: Spring–summer; white; fragrant; cup-shaped, 5–9 inches (1.3–2.3 dm) in diameter. FRUIT: Conelike receptacle, with numerous red seeds. RANGE: Southeastern U.S. NOTE: This elegant landscaping tree has several varieties. RELATED SPECIES: Pyramid magnolia (*M. pyramidata*), in Jasper and Newton counties, has large flowers and deciduous leaves that are broad in the middle and narrow at the base with 2 earlike appendages.

Sweetbay magnolia (*M. virginiana*), in swampy areas of East Texas, has flowers 2–3 inches (5–8 cm) wide and leaves that are deciduous to persistent, with the bottom pale or whitish, often velvety.

296. Rose Pavonia, *Pavonia lasiopetala*
   Malvaceae, Mallow Family

Small shrub to about 5 feet (1.5 m) tall, with slender branches; in dry rocky woods, usually in limestone; uncommon. LEAVES: Deciduous, simple, alternate; blade triangular, heart-shaped, or egg-shaped, ½–3 inches (1.2–7.5 cm) long; surface velvety; margins toothed and may have shallow lobes. FLOWERS: Nearly year-round; pink; to 1¾ inches (4.5 cm) wide, with 5 petals; stamen tube bending. FRUIT: Dry, seedlike. RANGE: Texas; Mexico. NOTE: This drought-tolerant plant is excellent for landscapes, growing rapidly from seed.

297. Chinaberry, *Melia azedarach*
   Meliaceae, Mahogany Family

Small to large tree to 75 feet (23 m) tall, with a broad rounded crown; widespread and common. LEAVES: Deciduous, alternate, twice pinnately compound; whole leaf may be 1–2 feet (3–6 dm) long; leaflets 1–3½ inches (2.5–9 cm) long; margins toothed, sometimes lobed. FLOWERS: Spring; lavender to white, fragrant; ½ inch (1.2 cm) wide; in showy panicles. FRUIT: Yellow drupes, ½–¾ inch (1.2–2 cm) in diameter, with single, ridged stone; poisonous. RANGE: Native of Asia, escaping cultivation in East and Central Texas.

298. Paper Mulberry, *Broussonetia papyrifera*
   Moraceae, Mulberry Family

Shrub or small tree to 27 feet (8.3 m) tall; sap milky; in disturbed ground; common. LEAVES: Deciduous, simple, alternate or opposite; blade oval, heart-, or egg-shaped, 3–9 inches (8–23 cm) long; top rough, bottom velvety; margins sharp-toothed, may be deeply lobed on 1 or both edges. FLOWERS: Spring; male flowers in catkins; female flowers in green round heads; dioecious. FRUIT: Red-orange fleshy wormlike fruits emerging from female flower-head; not edible. RANGE: Native of China and South Pacific, escaping from cultivation. NOTE: Natives of Samoa pound the inner bark into a papery cloth, known as tapa cloth. Female plants are rare in the U.S., so fruit is not usually seen.

299. Osage Orange, Bois d'Arc, Horse Apple, Naranjo Chino, *Maclura pomifera*
Moraceae, Mulberry Family

Medium-sized tree to 40 feet (12.3 m) tall, with milky sap, often with stout thorns on branches; orange on roots and beneath furrowed bark; widespread and common. LEAVES: Deciduous, simple, alternate; blade egg-shaped, lanceolate, or elliptic, 1–8 inches (2.5–20 cm) long; top shiny, often dark green; margins entire. FLOWERS: Spring; green, inconspicuous; dioecious. FRUIT: Green, wrinkled, resembling a brain, 3–6 inches (8–15 cm) in diameter; not edible. RANGE: Native to south-central U.S., escaping from cultivation outside natural range. NOTE: Osage orange is often planted as a hedgerow and windbreak. Native Americans used the wood for bows and the orange bark as a dye for wool.

300. Red Mulberry, Moral, *Morus rubra*
Moraceae, Mulberry Family

Medium-sized tree to 41 feet (12.6 m) tall, with milky sap; in moist woodlands; common. LEAVES: Deciduous, simple, alternate; blade broadly egg- to heart-shaped or deeply palmately lobed, 2–10 inches (5–25 cm) long; top surface rough or smooth and shiny; bottom with soft fuzzy hairs; margins toothed. FLOWERS: Spring; in catkins. FRUIT: Juicy, 1¼ inch (3 cm) long, blackberry-type, purple-black when ripe; edible. RANGE: Eastern half of U.S. RELATED SPECIES: White mulberry (*Morus alba*), a Chinese ornamental, has leaves with the top shiny and smooth to slightly rough, the bottom not fuzzy (sometimes with hairs); varieties include red, white, and black fruits and fruitless trees. Texas mulberry (*M. microphylla*), in canyons and hills of West Texas, has small leaves that are sandpapery rough on both surfaces; the fruit is small. NOTE: The pollen is allergenic.

301. Wax Myrtle, Bayberry, *Morella cerifera*
(Myrica cerifera)
Myricaceae, Bayberry Family

Fragrant shrub or small tree to 22 feet (5.3 m) tall; understory shrub, along streams and reservoirs, in wetlands; common. LEAVES: Evergreen, simple, alternate; blade narrow, elliptic or oblanceolate, 1–5 inches (2.5–13 cm) long, usually less than ¾ inch (2 cm) wide; bayberry aroma; margins entire or toothed in upper half. FLOWERS: Spring; in catkins. FRUIT: Hard round fruits ⅛ inch (4 mm) in diameter; coated with white wax. RANGE: Eastern U.S.; Central America. NOTE: The leaves are used as a spice, like bay leaves. The wax of the fruit has been used to make bayberry candles; popular landscape plant. RELATED SPECIES: Hybridizes with *M. caroliniensis*, a shrub usually less than 10 feet (3 m) tall, with blades wider than ¾ inch (2 cm).

302. Fringe Tree,
Old Man's Beard, *Chionanthus virginicus*
Oleaceae, Olive Family

Shrub or small tree to 27 feet (8 m) tall; in moist woods, wetlands, thickets; uncommon. LEAVES: Deciduous to persistent, simple, alternate or opposite; blade usually elliptic, 4–8 inches (1–2 dm) long; bottom pale, may be hairy; margins entire. FLOWERS: Early spring before and as leaves emerge; white or greenish; fragrant; petals linear, tassellike, 1–2 inches (2.5–5 cm) long; dioecious. FRUIT: Purple drupe ⅜–1 inch (1–2.5 cm) long. RANGE: Eastern U.S.

303. Narrow-leaf Forestiera, Desert Olive, Panalero, *Forestiera angustifolia*
Oleaceae, Olive Family

Shrub or small tree; branches slender; in brush; common. LEAVES: Persistent, simple, opposite or clusters opposite; blade linear to narrowly oblanceolate, ½–1¼ inches (1–3 cm) long; margins entire. FLOWERS: Late winter–spring, usually before leaves; greenish, no petals. FRUIT: Purple to black drupe, to ½ inch (1.2 cm) long, slightly curved. RANGE: Texas; Mexico.

304. Spring Herald, Elbow Bush, Cruzilla, *Forestiera pubescens*
Oleaceae, Olive Family

Shrub about 3–15 feet (1–4.5 m) tall, with slender curving branches; along streams, in brush, prairies, fields; common. LEAVES: Deciduous, simple, opposite; blade broad, elliptic, oval, or egg-shaped, ½–2 inches (1.5–5 cm) long; surfaces smooth or hairy; margins finely toothed. FLOWERS: Late winter–early spring, before or with emerging leaves; greenish yellow, no petals; usually dioecious. FRUIT: Blue-black fleshy drupe, oblong or oval, to ⅜ inch (1 cm) long. RANGE: Oklahoma to California, with populations in South Carolina and Florida; Mexico. RELATED SPECIES: Swamp privet, *F. acuminata*, in wetlands in the eastern half of Texas to Dallas County, has blades usually longer than 1¼ inches (3.5 cm) and spring flowers; *F. ligustrina*, in sandy soil in East Texas, has flowers late summer–fall; *F. reticulata*, on dry limestone hills and canyons of Central Texas, has shiny, persistent leaves and summer flowers.

(photos 305, 306, 307) Ash, *Fraxinus* species (see Key)
Oleaceae, Olive Family

Shrubs or trees with paired branches. LEAVES: Deciduous in most species, opposite, pinnately compound; leaflets toothed or entire. FLOWERS: Usually inconspicuous (except *F. cuspidata*, with showy white flowers); usually dioecious. FRUIT: Elongated samara (a winged seed), in some species resembling a canoe paddle. NOTE: Often used in landscaping.

KEY TO ASHES

1a. East of the Trans-Pecos .... go to 4

1b. In the Trans-Pecos, east to Val Verde County .... go to 2

2a (1b). Leaflets small, ¼–1¼ inches (7–35 mm) long, to ⅜ inch (1 cm) wide; often only 3 leaflets per leaf (but can have 5 or 7); midrib usually winged; leaves leathery, persistent; samara oblong, to ¼ inch (6 mm) wide; shrub to small tree; in mountains, canyons, rocky hills near the Rio Grande ....
.... Gregg ash, littleleaf ash, escobilla, *Fraxinus greggii*

2b. Leaflets longer than 1 inch (2.5 cm); deciduous, usually thin; 3–9 leaflets per leaf .... go to 3

3a (2b). Flowers white, fragrant, showy, ½ inch (13 mm) long, with 4 linear lobes; usually with 7 leaflets (can have 3–9), 1¼–2½ inches (3.5–7 cm) long, tip long-pointed; top surface shiny, dark green; samara with wing surrounding part or most of flattened seed; shrub or small tree; in mountains and canyons ....
.... fragrant ash, *Fraxinum cuspidata* (photo 307)

3b. Flowers inconspicuous, appearing before leaves; usually with 5 leaflets (can have 3–9); leafless variable in size and shape, 1–4 inches (2.5–10 cm) long, from thin to leathery and from hairless to densely woolly underneath, tapering or not; samara with wing narrowly surrounding up to half of cylindrical seed ....
Arizona ash, velvet ash, *Fraxinus velutina*—Small to large tree; along streams, in canyons, mountains, common; popular in landscaping; also see *F. berlandieriani* (5a.).

4a (1a). Leaflets broad, elliptic or egg-shaped to nearly round (rarely narrow), 1–3 (or 4) inches (2.5–10 cm) long; on dry rocky hills, along streams and reservoirs, endemic to Central and North Central Texas along the Balcones Escarpment ....

Texas ash, *Fraxinus texensis* (photo 306)—Small to medium tree; 5 or 7 leaflets, margins entire or teeth usually rounded. Samara ⅜–1¼ inch (1.5–3 cm) long with wing attached to 1 end of cylindrical seed; seed body to ⅝ inch (1.5 cm) long.

4b. Leaflets longer than 2 inches (5 cm), typically 2–3 times longer than wide; in moist or well-drained soil, in woods, near streams, in swamps . . . . go to 5

5a (4b). In South Texas, in Bastrop and Travis counties, and along canyons of the Rio Grande from Boquillas Canyon east . . . .
Mexican ash, Fresno, *Fraxinus berlandieriani*—Small to medium tree; 3 or 5 leaflets, 2–5 inches (5–12.5 cm) long; samaras about 1–1½ inches (2.5–4 cm) long with wing narrowed at base, surrounding part or all of seed; popular in landscaping; mistakenly called Arizona ash.

5b. In eastern half of Texas or north to Panhandle; samaras typically 1–3 inches (2.5–8 cm) long . . . . go to 6

6a (5b). Found only in deep Southeast Texas, in swamps and along streams; broad wing of samara ⅜–1 inch (1–2.5 cm) wide, completely surrounding seed; 5 or 7 leaflets to 5 inches (12.5 cm) long . . . .
. . . . Carolina ash, *Fraxinus caroliniana*

6b. Not restricted to Southeast Texas; narrow wing of samara not usually surrounding entire seed . . . . one of the following:

(i) Green ash, red ash, *Fraxinus pensylvanica*—-Large tree, in eastern half of Texas and in northern Panhandle; leaflets usually 7 (can have 5 or 9), 2–6 inches (5–15 cm) long, often less than 1 inch (2.5 cm) wide but can be wider; leaf top bright green, bottom lighter green but not whitish; wing of samara attached to 1 end of seed or nearly surrounding seed; seed body to 1¼ inch (3 cm) long.

(ii) White ash, *Fraxinus americana* (photo 305)—Large tree in East Texas; usually with 7 leaflets (can have 5 or 9); leaflets 2–6 inches (5–15 cm) long, to 3 inches (7.5 cm) wide, top dark green and bottom pale, glaucous, or whitish; wing of samara attached to end of seed; seed body to ⅝ inch (1.5 cm) long.

308. Ligustrum, Japanese Privet, *Ligustrum japonicum*
Oleaceae, Olive Family

Shrub or tree to 35 feet (10 m) tall; often forming thickets; abundant. LEAVES: Evergreen in mild winters, simple, opposite; blade egg-shaped to oval, 2–6 inches (5–15 cm) long; top shiny, smooth, not hairy, dark green; margins entire. FLOWERS: Late spring–summer; white; fragrant; tiny; in showy panicles. FRUIT: Blue-black, ¼ inch (6 mm) oval drupe; deadly poisonous. RANGE: Asian native, escaping cultivation in eastern half of Texas. RELATED SPECIES: Numerous varieties and several other ornamental species cultivated in Texas.

(photos 309, 310) Pine, *Pinus* species (see Key)
Pinaceae, Pine Family

Evergreen cone-bearing trees; sap aromatic. LEAVES: needlelike, in bundles (fascicles) of 2 or more. NOTE: Pine nuts, particularly the large nuts of the pinyons, are edible. East Texas pines have great economic value for lumber and pulp. Most of East Texas' pine-hardwood forests have been clear-cut, to be replaced with pine monoculture or residential or industrial areas. SIMILAR SPECIES: Blue Douglas fir (*Pseudotsuga menziesii*), in the Chisos and Guadalupe mountains, has short needles, ¼–1¼ inch (6–30 mm) long, that occur singly, not in bundles. See bald cypress, page 276. Fir and spruce trees, also with solitary needles, occur in Texas only as ornamentals. Bark used as a dye.

KEY TO PINES

1a. In West Texas . . . . go to 5

1b. In East Texas, west to Bastrop County . . . . go to 2

2a (1b). Needles about 2–6 inches (5–15 cm) long, 2 or 3 per bundle; cones small, to 2¾ inches (7 cm) long; in East Texas to Bryan area . . . . shortleaf yellow pine, *Pinus echinata*

2b. Needles usually longer than 5 inches (1.2 dm); cones longer than 2 inches (5 cm) . . . . go to 3

3a (2b). Needles 6–18 inches (15–45 cm) long, 3 per bundle; cones generally 6–10 inches (1.5–2.5 dm) long; in Southeast Texas . . . . longleaf yellow pine, *Pinus palustris*

3b. Needles 5–11 inches (1.2–2.8 dm) long, 2 or 3 per bundle; cones less than 6 inches (1.5 dm) long . . . . go to 4

4a (3b). Extensively cultivated in East Texas (may also be naturalized); cones on short stalk . . . . slash pine, *Pinus elliottii*

4b. Native to East Texas, west to Bastrop County; cones 1½–5 inches (4–13 cm) long, without stalk .... loblolly pine, *Pinus taeda* (photo 310)

5a (1a). Needles 5 per bundle, 1½–4 inches (4–10 cm) long; cones 3–10 inches (8–25 cm) long; in Guadalupe and Davis mountains .... southwestern white pine, limber pine, *Pinus strobiformis*

5b. Needles generally 2 or 3 per bundle, needles short or long (if 4 or 5 per bundle, needles longer than 3 inches, or 8 cm) .... go to 6

6a (5b). Small to large tree; needles 1–2¼ inches (2.5–5.7 cm) long; cones small, 1–2¼ inches (2.5–6 cm) long .... go to 8 (pinyons)

6b. Large tree; needles 3–12 inches (1–3 dm) long; cones 2–6 inches (5–15 cm) long .... go to 7

7a (6b). In Guadalupe and Davis mountains; needles 2 or 3 per bundle, rarely 4 or 5; sap smells like vanilla .... ponderosa pine, *Pinus ponderosa*

7b. Rare in the Chisos Mountains; needles 2 or 3 per bundle, occasionally 4 or 5 .... *Pinus arizonica* var. *stormiae*

8a (6a). In mountains of Culberson and Hudspeth counties (and possibly in Panhandle in Deaf Smith County) .... Colorado pinyon, *Pinus edulis* (photo 309)

8b. In Southern Trans-Pecos or on Edwards Plateau .... go to 9

9a (8b). Trans-Pecos and possibly Edwards Plateau, usually in igneous rocky soil above 5,000 feet elevation; shells of nuts hard .... Mexican pinyon, *Pinus cembroides*

9b. In scattered, isolated populations from southern Trans-Pecos to western Edwards Plateau, usually in limestone rocky soil below 5,000 feet; shells of nuts thin; needles usually 2 (can be 1 or 3) per bundle .... papershell pinyon, *Pinus remota*

311. Sycamore, Plane Tree, *Platanus occidentalis*
Platanaceae, Sycamore Family

Tree 40–160 feet (12–50 m) tall; tallest deciduous tree in U.S. (state champion 90 feet, or 27.5 m); older bark peels to expose smooth white or creamy bark; in bottomlands, along streams, lakes, swamps; common, with several varieties. LEAVES: Deciduous, simple, alternate; blade very broad, 4–12 inches (1–3 dm) wide; smooth or velvety; margin palmately lobed, usually with 3–5 lobes; tips of lobes pointed, often tapering. FLOWERS: Spring; inconspicuous

green balls. FRUIT: Brown ball, 1¼ inches (3 cm) in diameter, dangling from long stem. RANGE: Eastern U.S.; Canada. NOTE: This popular ornamental is often cultivated outside natural range. Bark and leaves yield nice dyes.

(photos 312, 313) Redroot, New Jersey Tea,
Ceanothus americanus
Rhamnaceae, Buckthorn Family

Delicate shrub to about 3 feet (1 m) tall, in woods, prairies; uncommon. LEAVES: Deciduous, simple, alternate; blade broad, elliptic to egg-shaped, 2–4 inches (5–10 cm) long; usually hairy; margins finely toothed. FLOWERS: Spring; white; tiny; in showy rounded clusters about 1 inch (2.5 cm) long. FRUIT: Small 3-lobed capsule. RANGE: Eastern U.S.; Canada. NOTE: Scalded and dried, the leaves were a popular tea substitute during the American Revolution. RELATED SPECIES: *Ceanothus herbaceus*, from Southeast Texas to the Panhandle, has narrow leaves to 2½ inches (6.5 cm) long, with few or no hairs. Two species in Trans-Pecos mountains and canyons are 2–6 feet (6–20 dm) tall, with narrowly elliptic leaves less than 1 inch (2.5 cm) long; *C. fendleri*, rare at high elevations, with thorn-tipped branches and margins very finely toothed to entire; and *C. greggii* (photo 313), common, with leaves opposite, margins usually entire, leaf bottom dotted with patches of white hairs (hand lens).

314. Snakewood, *Colubrina texensis*
Rhamnaceae, Buckthorn Family

Thicket-forming shrub about 3–6 feet (1–2 m) tall; gray bark; branches zigzagging, without thorns; in arid brush; common. LEAVES: Deciduous, simple, alternate or clustered; blade elliptic, oblanceolate, or egg-shaped; ½–1½ inches (1.2–4 cm) long; margins finely toothed. FLOWERS: Early spring–summer; green, inconspicuous nectar-filled disk. FRUIT: Hard drupe ⅜ inch (1 cm) in diameter; thin skin turning black; 2–3 seeds. RANGE: Texas; Mexico.

315. Javelina Bush, Tecomblate, *Condalia ericoides*
Rhamnaceae, Buckthorn Family

Small densely branched shrub, typically less than 3 feet (1 m) tall; branches tipped with thorns; in brush, deserts, mountains; widespread and common. LEAVES: Evergreen, simple, alternate or in clusters; blade linear, ⅛–½ inch (3–13 mm) long, to ½₂ inch (1 mm) wide; margins entire, edges grooved on leaf bottom. FLOWERS: Spring–summer; yellow; inconspicuous. FRUIT: Reddish purple to black drupe, oblong, ⅛–½ inch (4–12 mm) long. RANGE: Texas to Arizona; Mexico. SIMILAR SPECIES: See brasil, below. *Condalia warnockii*, with leaves to ¼ inch (6 mm) long, to ⅛ inch (2.6 mm) wide, widest near tip, has a similar range as *C. ericoides*. Amargoso (*Castela erecta* ssp. *texana*), an unrelated small shrub in South Texas, west to Terrell County and north to Travis County, has linear to narrowly lanceolate leaves to 1 inch (2.5 cm) long, red to orange flowers, and red nearly circular, flattened fruit.

316. Brasil, Bluewood, Capul Negro, *Condalia hookeri*
Rhamnaceae, Buckthorn Family

Thicket-forming shrub or small tree to 30 feet (9.2 m) tall; branches may be tipped with thorns; in dry brush or woods; widespread and common, with 2 varieties. LEAVES: Persistent, simple, alternate or clustered; blade paddle-shaped, widest at tip, narrow at base, ⅜–1½ inches (1–4 cm) long; margins entire. FLOWERS: Spring–summer; green; inconspicuous. FRUIT: Fleshy blue-black drupe ¼ inch (6 mm) in diameter; edible. RANGE: Texas; Mexico. NOTE: Var. *edwardsiana*, in Edwards County, is in danger of extinction. SIMILAR SPECIES: 2 species with leaves typically less than ½ inch (1.2 cm) long, to ³⁄₁₆ inch (4 mm) wide, widest near tip—

C. spathulata, typically less than 3 feet (1 m) tall, in South Texas to the southern Edwards Plateau; *C. viridis*, shrub in the Trans-Pecos and western Edwards Plateau. NOTE: Fruits make a nice dye.

317. Coyotillo, *Karwinskia humboldtiana*
Rhamnaceae, Buckthorn Family

Rounded shrub about 3–6 feet (1–2 m) tall; in dry brush; common. LEAVES: Deciduous, simple, opposite; blade oblong, elliptic, or lanceolate, ¾–3 inches (2–7.5 cm) long; veins conspicuous, parallel; margins entire or with a few rounded teeth. FLOWERS: Summer–fall; greenish; inconspicuous. FRUIT: Fleshy black drupe to ⅜ inch (1 cm) in diameter; poisonous. RANGE: Texas; Mexico.

318. Carolina Buckthorn, Indian Cherry,
Rhamnus caroliniana (Frangula caroliniana)
Rhamnaceae, Buckthorn Family

Shrub or small tree to about 40 feet (12 m) tall; in moist woods and bottomlands, along streams; common. LEAVES: Deciduous or persistent, simple, alternate; blade oblong to elliptic, 2–6 inches (5–15 cm) long; veins parallel, prominent on bottom; top shiny, smooth, usually nearly hairless (bottom smooth or occasionally velvety); margins entire or finely toothed. FLOWERS: Spring–summer; green; inconspicuous. FRUIT: Red to black fleshy fruit about ⅜ inch (1 cm) in diameter; not edible, poisonous in some species. RANGE: Eastern U.S. south of Virginia. RELATED SPECIES: 2 species in Trans-Pecos mountains, often with leaf bottom velvety—*R. betulifolia*, with leaves 2–6 inches (5–15 cm) long; and *R. serrata*, a small shrub less than 9 feet (3 m) tall, with leaves less than 2 inches (5 cm) long. *Rhamnus lanceolata*, in East Texas, has flowers before or with emerging leaves, leaves to 3½ inches (9 cm) long, and veins not prominent.

319. Lotebush, *Ziziphus obtusifolia*
Rhamnaceae, Buckthorn Family

Thicket-forming shrub about 3–6 feet (1–2 m) tall; branches rigid, tipped with thorns; new growth green or gray-green, often with whitish waxy coating; in various habitats; common and widespread. LEAVES: Deciduous, simple, alternate or clustered; blade ½–1½ inches (1.2–4 cm) long; shape variable, oblong to almost linear, heart-shaped, or egg-shaped; margins entire or with small teeth. FLOWERS: Spring–summer; green; inconspicuous. FRUIT: Fleshy blue-black drupe ⅜ inch (1 cm) in diameter, with a whitish waxy coating; edible. RANGE: Southwestern U.S.; Mexico. RELATED SPECIES: Jujube (*Z. jujuba*), an Asian ornamental tree, has paired thorns at leaf bases and along the zigzag branches (there is also a thornless variety); the fleshy yellow, reddish brown, or black fruit is to 1 inch (2.5 cm) long, sweet and edible, and resembles a date.

320. Mountain Mahogany, *Cercocarpus montanus*
Rosaceae, Rose Family

Shrub or small tree about 3–15 feet (1–4.5 m) tall; in open woods, rocky hills, canyons, mountains; widespread, with 3 varieties. LEAVES: Evergreen or persistent, simple, alternate or clustered; blade oblong, oval, elliptic or obovate; size variable, either with leaves minute (less than ¾ inch, 2 cm) and margins mostly entire, or with leaves to 2½ inches (6.5 cm) long, toothed, usually just in upper half; surface hairy or not; veins prominent on bottom. FLOWERS: Spring–summer; creamy white or yellowish; minute, inconspicuous. FRUIT: Single achene attached to a feathery corkscrew-shaped tail 1–3 inches (2.5–8 cm) long. RANGE: Western U.S.; Mexico.

321. Hawthorn, Thornapple, *Crataegus texana*
    Rosaceae, Rose Family

Small to medium-sized tree to about 30 feet (9 m) tall; branches armed with stout straight thorns or unarmed; in moist woods and thickets, along streams; uncommon. LEAVES: Deciduous, simple, alternate; blade egg-shaped, triangular, or oval, 1–4 inches (2.5–10 cm) long and nearly as wide; mature leaf shiny on top, densely hairy to felty on bottom; margins sharp-toothed, sometimes lobed. FLOWERS: Spring; white; to 1 inch (2.5 cm) across; often with red or dark center; 5 petals, numerous stamens. FRUIT: Fall; sweet, red, ½–1 inch (1.2–2.5 cm) in diameter; good for jelly. RANGE: Endemic to Texas. RELATED SPECIES: *Crataegus mollis* may be merged with this species. 17 Texas species difficult to distinguish. Several endemic species.

322. Apache Plume, *Fallugia paradoxa*
    Rosaceae, Rose Family

Intricately branched shrub to about 8 feet (2.5 m) tall; conspicuous when covered with feathery fruits; in dry sunny rocky hills, mountains; common. LEAVES: Deciduous or persistent, simple, alternate or clustered; blade ¼–1 inch (6-25 mm) long; margins with 3-7 fingerlike lobes. FLOWERS: Spring–fall; white; to 1½ inches (4 cm) wide, 5 or 6 petals. FRUIT: Cluster of achenes attached to feathery tails 1–2 inches (2.5–5 cm) long; resembles tiny feather duster. RANGE: Texas to California; Mexico. SIMILAR SPECIES: Heath cliffrose (*Purshia ericifolia*), a low shrub in the Trans-Pecos, has similar feathery fruits, but its leaves are unlobed, linear and spiny, to ¼ inch (6 mm) long.

(photos 323, 324) Cherry, Chokecherry, *Prunus* **species (see key)**
    Rosaceae, Rose Family

Trees or shrubs. LEAVES: Simple, alternate; blade oblong, elliptic, lanceolate, or egg-shaped. FLOWERS: Late winter-spring;

white; fragrant; to ½ inch (12 mm) wide; in elongated racemes. FRUIT: Dark red to black drupe, fleshy or dry, to ½ inch (1.2 cm) in diameter; in elongated racemes. NOTE: The seeds, leaves, and bark contain toxic cyanide-forming glycosides.

KEY TO CHERRIES

1a. Leaf margins entire or with spine-tipped teeth; leaves evergreen; blade 2–4½ inches (5–12 cm) long; small tree in East Texas, mainly east of Houston . . . .
    Laurel cherry, *Prunus caroliniana* (photo 323)—racemes to 1½ inches (4 cm) long; fruit dry, astringent, not edible; uncommon but often used in landscaping.

1b. Leaf margins with fine or coarse teeth, but teeth not spine-tipped; leaves deciduous or persistent; blade 1½–6 inches (4–15 cm) long, often with hairs along midrib; shrub or small to large tree with 3 subspecies, in East Texas, on Edwards Plateau, and in the Trans-Pecos . . . .
    Black cherry, southwestern chokecherry, *Prunus serotina* (photo 324)—Uncommon; bark shiny, with light-colored lenticels (short corky horizontal lines); older bark dark, flaky; flowers in racemes 1–6 inches (2.5–15 cm) long; sepal tips persistent on fruit; fruit edible when cooked, but some too bitter for use. RELATED SPECIES: Common chokecherry (*P. virginiana*), a shrub or small tree in the Panhandle and Trans-Pecos, has leaf margins more finely toothed and sepals not persistent on fruit.

325. Plum, *Prunus* species (see Key)
Rosaceae, Rose Family

Shrubs and small trees; bark often with light-colored lenticels (short corky horizontal lines); young twigs reddish brown or gray; branches of some species spine-tipped; most species scattered, not common. LEAVES: Deciduous, simple, alternate or clustered on stubby spurs; blades oblong, elliptic, oval, lanceolate, or egg-shaped. FLOWERS: Late winter–spring; in some species, covering branches before leaves emerge; white; fragrant; small, ⅜–1 inch (1–2.5 cm) wide; 5 petals, many stamens; solitary, or flower stems joined to form small rounded or flat-topped clusters. FRUIT: Fleshy drupe; red, purple, or yellow; elliptic or round; edible but some too sour to eat uncooked. NOTE: The seeds, leaves, and bark contain toxic cyanide-forming glycosides. RELATED SPECIES: 4 dry-fruited species, rare or restricted in range, are not included in the key. Non-native ornamental and fruit trees are not included. Hybrids create difficulties in identifying the native plums.

KEY TO PLUMS

1a. Top of leaf dull (not shiny), may be densely hairy or not, smooth, or rough; bottom of mature leaf densely hairy, or hairs mainly along veins (hand lens) . . . . go to 3

1b. Top of leaf smooth and usually shiny; bottom of mature leaf with or without hairs, but hairs not dense (may be restricted to midrib) . . . . go to 2

2a (1b). Leaves narrow, 2–3 times longer than broad; blade 1–3 inches (2.5–8 cm) long, ⅜–⅞ inch (1–2.2 cm) wide; branches often spine-tipped . . . .
Chickasaw plum, *Prunus angustifolia* (photo 325)—thicket-forming shrub or small tree; from East Texas through Central Texas to Panhandle; leaf margins finely toothed; fruit ½–¾ inch (1.2–2 cm) long.

2b. Leaves about 2 times longer than broad . . . . one of the following:

    (i) Flatwoods plum, *Prunus umbellata*—Shrub or small tree, not thicket-forming; in East Texas; top of leaf dull, with or without hairs, hairs on bottom may be restricted to midrub; fruit ⅜–¾ inch (1–2 cm) in diameter.

    (ii) Creek plum, *Prunus rivularis*—Small thicket-forming shrub; in Edwards Plateau and North Central Texas; leaf blades 2–3 inches (5–7 cm) long, margins finely toothed; fruit ½–¾ inch (1.2–2 cm) long.

    (iii) Wild goose plum, *Prunus munsoniana*—very rare thicket-forming shrub or small tree; in isolated populations in North Central Texas, Central Texas, and along the coast; leaf blade broad, 1½–4 inches (4–10 cm) long, margins finely toothed; fruit ½–1 inch (1.2–2.5 cm) long.

3a (1a). Small shrub, often thicket-forming; leaves about 1–2 inches (2.5–5 cm) long, ½–1 inch (1.2–2.5 cm) wide . . . .
Oklahoma plum, *Prunus gracilis*—In East and North Central Texas and Panhandle; leaf margins finely toothed; fruit ½–¾ inch (1.2–2 cm) long.

3b. Shrub or small tree, not thicket-forming; leaves usually longer than 1½
    inches (4 cm) .... go to 4

4a (3b). Leaves narrow or broad, usually shorter than 2¾ inches (7 cm);
    margins finely toothed ....
    .... flatwoods plum, *Prunus umbellata* (see 2b.)

4b. Leaves broad, usually longer than 2½ inches (6.5 cm), can be 2–5 inches
    (5–13 cm) long; margins with fairly coarse teeth ....
    Mexican plum, *Prunus mexicana*—to 26 feet (7.9 m) tall
    (national champion), with broad crown; in East Texas from
    northeast corner to eastern Edwards Plateau; fruit ½–1¼ inches
    (1.2–3 cm) in diameter.

326. Scarlet Bouvardia, Trompetilla
Bouvardia ternifolia
Rubiaceae, Madder Family

Shrub to about 4 feet (1.2 m) tall, with slender upright branches;
in canyons, on rocky slopes and mountains; uncommon. LEAVES:
Deciduous, simple, opposite or whorled; blade narrow, lanceolate
to egg-shaped, 1–2 inches (2.5–5 cm) long; surface may be rough;
margins entire. FLOWERS: Late spring–fall; flamboyant red;
4-lobed, tubular, ¾–1½ inches (2–4 cm) long. FRUIT: Small capsule
with numerous tiny seeds. RANGE: Texas to Arizona; Mexico.
NOTE: This shrub would be a beautiful landscape plant.

327. Buttonbush, *Cephalanthus occidentalis*
Rubiaceae, Madder Family

Shrub or small tree 5 to 13 feet (1.5–4 m) tall; in moist soil, along
streams, in swamps and wet areas; widespread and common.
LEAVES: Deciduous, simple, opposite or whorled; blade elliptic to
lanceolate, 2–8 inches (5–20 cm) long; margins entire to slightly
wavy. FLOWERS: Spring–summer; white; fragrant; in showy round
heads 1 inch (2.5 cm) in diameter. FRUIT: Dense round head composed of many brown capsules. RANGE: Throughout most of U.S.

328. Hop Tree, Wafer Ash, Skunkbush, Cola de Zorrillo, *Ptelea trifoliata* Rutaceae, Citrus Family

Shrub about 4–8 feet (1.2–2.5 m) tall, or occasionally a small tree; in forests, rocky hills, canyons, mountains, sand dunes; widespread, with numerous subspecies and varieties. LEAVES: Deciduous, trifoliate, alternate; leaflets variable in size and shape, ½–4 inches (1.2–10 cm) long; strongly aromatic; bottom smooth to velvety; margins entire or finely toothed. FLOWERS: Spring–summer; inconspicuous; greenish or yellowish white. FRUIT: Thin waferlike round samara (winged seed) to 1 inch (2.5 cm) in diameter. RANGE: Much of U.S.; Canada; Mexico.

(photos 329, 330) Prickly Ash, Hercules' Club, Toothache Tree, Tickle-tongue, *Zanthoxylum* species (see Key) Rutaceae, Citrus Family

Shrubs or small trees with curved or straight spines on branches and trunk, sometimes on leaves; leaves and bark aromatic. LEAVES: Alternate, pinnately compound; margins toothed. FLOWERS: Greenish; tiny; in many-flowered clusters; usually dioecious. FRUIT: Red or brown leathery capsule, splitting open to reveal a black shiny seed. NOTE: The leaves of most species cause numbness in the mouth when chewed.

KEY TO PRICKLY ASHES

1a. In Davis Mountains and Brewster County; 7–9 leaflets; leaflets often less than ½ inch (1.2 cm) long, can be to 1½ inches (4 cm); endemic and endangered . . . . Shinners' tickle-tongue, *Zanthoxylum parvum*

1b. Not in Trans-Pecos . . . . go to 2

2a (1b). Midrib of compound leaf winged; leaves persistent; 5–19 leaflets, each ¼–1½ inches (0.7–4 cm) long; flower clusters small, inconspicuous; common in thickets and woods; in South Texas and along coast to Galveston . . . . lime prickly ash, colima, *Zanthoxylum fagara* (photo 329)

2b. Midrib not winged; leaves deciduous, strongly aromatic, top surface often shiny . . . . go to 3

3a (2b). Leaflets fairly large, ½–4½ inches (1.2–11.5 cm) long; 5–19 leaflets; flower clusters conspicuous, 1–6 inches (2.5–15 cm) long; spines on trunk often emerging from conical bases; mainly in eastern third of Texas; in openings in woodlands, along fencerows; uncommon . . . Hercules' club, prickly ash, *Zanthoxylum clava-herculis* (photo 330)

3b. Leaflets smaller, usually ¼–1¾ inches (0.6–4.5 cm) long; 3–11 leaflets; some populations with leaves and twigs densely hairy; flower clusters small, inconspicuous; in South and Central to North Central Texas; uncommon . . . . prickly ash, toothache tree, tickle-tongue, *Zanthoxylum hirsutum*

331. Eastern Cottonwood, Alamo, *Populus deltoides*
Salicaceae, Willow Family

   Large, fast-growing tree to 110 feet (33.6 m) tall, often with straight trunk and furrowed bark; near water, along streams, ponds, roadsides; common, with 3 subspecies: *deltoides*, in eastern half of Texas; *monilifera*, in Panhandle; *wislizenii*, in Trans-Pecos. LEAVES: Deciduous, simple, alternate; blade triangular to egg-shaped, 3–7 inches (7.5–18 cm) long and wide; margins with large teeth; stems long, flattened. FLOWERS: Spring; in long catkins; dioecious. FRUIT: Tiny round capsules, numerous on long, pendant stalk; seeds attached to cotton fluff. RANGE: U.S. and Canada east of Rocky Mountains. RELATED SPECIES: *Populus fremontii*, in the Trans-Pecos, has diamond-shaped leaves and is not always easily distinguished; quaking aspen, (*P. tremuloides*), only in high mountains above 7,000 feet, has white bark and small, nearly round leaves.

332. Black Willow, *Salix nigra* (includes *S. goodingii*)
Salicaceae, Willow Family

   Medium to large tree to 89 feet (27.2 m) tall, trunk often bending; bark on older trees dark, deeply ridged, and shaggy; along streams, other wet areas; widespread and common, with 2 varieties.

LEAVES: Deciduous, simple, alternate; blades linear to narrowly lanceolate, 2–7 inches (5–18 cm) long; margins finely toothed. FLOWERS: Spring; in fragrant catkins; dioecious. FRUIT: Tiny capsule; seeds attached to silky hairs. RANGE: Eastern U.S.; Canada; Mexico. NOTE: The slender, supple branches are useful for weaving baskets; leaves and bark make nice dyes. RELATED SPECIES: 7 other species in Texas. Weeping willow (*Salix babylonica*), a Chinese ornamental, has long, drooping branches.

333. Western Soapberry, Jaboncillo,
   Sapindus saponaria var. *drummondii*
   Sapindaceae, Soapberry Family

Small to large tree to 54 feet (16.5 m) tall; along streams and edges of woods, in fields; bark on older trees often shaggy; common. LEAVES: Deciduous, turning yellow in fall; alternate, pinnately compound; leaflets numerous, lanceolate, often falcate, 1½–4 inches (4–10 cm) long, often not paired; margins entire. FLOWERS: Spring–summer; white; tiny; in showy erect panicles; dioecious. FRUIT: Amber-colored fleshy translucent drupe to about ½ inch (1.5 cm) in diameter. RANGE: Louisiana to Arizona, north to Kansas; Mexico. NOTE: The fruit is toxic if eaten but can be used as a laundry soap and yields a nice dye. SIMILAR SPECIES: Flameleaf sumac (page 173) has red fruit, and its leaves turn red.

334. Mexican Buckeye, Monilla, *Ungnadia speciosa*
   Sapindaceae, Soapberry Family

Shrub or small tree, often with several slender trunks, 10–32 feet (3–10 m) tall; on wooded rocky slopes and canyons, near streams; common. LEAVES: Deciduous, alternate, pinnately compound, with 3–7 leaflets, leaflets lanceolate to egg-shaped, 3–7 inches (8–18 cm) long; margins toothed. FLOWERS: Emerging with or before leaves; pink to purple; fragrant; to 1 inch (2.5 cm) across. FRUIT: a 3-lobed capsule, 1–2 inches (2.5–5 cm) wide, green to cinnamon brown, woody or leathery; seeds hard, shiny, dark brown to black with pale scar, to ⅝ inch (1.5 cm) in diameter. RANGE: Texas, New Mexico; Mexico. NOTE: The seeds are poisonous. *Ungnadia* is not related to the buckeyes (*Aesculus* species, page 248).

335. Coma, Gum Elastic, Chittimwood,
Sideroxylon lanuginosum (Bumelia lanuginosa)
Sapotaceae, Sapodilla Family

Small shrub to large tree 6–80 feet (2–24.4 m) tall; branches often spine-tipped; sap may be milky; along streams, in woods, thickets, fields; widespread and common, with 5 varieties. LEAVES: Deciduous or persistent, simple, alternate or clustered; blade elliptic to oblanceolate, 1–4 inches (2.5–10 cm) long; bottom hairy, often woolly (1 variety hairless); margins entire. FLOWERS: Spring–summer; tiny; inconspicuous. FRUIT: Purple–black fleshy oval drupe ¼–1 inch (7–25 mm) long; sweet and suitable for jelly; raw fruit may cause stomach disturbance if eaten in quantity. RANGE: Southern U.S. from Arizona to Florida; Mexico. RELATED SPECIES: *S. celastrinum*, in thickets and salt marshes in South Texas, has spines along the branches and has evergreen leaves that are usually less than 1½ inches (4 cm) long, and usually hairless. *S. lycioides*, rare in Southeast Texas, has leaves that are hairless or, when young, have bottoms with silvery hairs.

336. Cenizo, Texas Silverleaf,
Purple Sage, *Leucophyllum frutescens*
Scrophulariaceae, Figwort Family

Shrub about 3–10 feet (1–3 m) tall; white hairs dense on leaves and tips of twigs; in rocky or sandy soil, in brush, canyons, dry plains, mountains; locally abundant. LEAVES: Evergreen, simple, alternate or clustered; blade elliptic to egg-shaped, ½–1 inch (1.2–2.5 cm) long;

surface gray-green; margins entire. FLOWERS: year-round; purple (with a white variety); to 1 inch (2.5 cm) wide; 5-lobed, bell-shaped. RANGE: Texas, New Mexico; Mexico. NOTE: This popular landscape plant has several varieties, including ones with silver leaves and green leaves. RELATED SPECIES: 2 shrubs typically less than 3 feet (1 m) tall, with smaller white or silver-gray leaves and smaller flowers, and veins on leaf bottoms obscured by woolly hairs—*Leucophyllum candidum*, in Brewster County; *L. minus*, western Edwards Plateau through the Trans-Pecos, with branches somewhat spine-tipped.

337. Tree of Heaven, *Ailanthus altissima*
Simaroubaceae, Quassia Family

Medium to large rapid-growing tree to 40 feet (12.2 m) tall; leaf scar large, shield-shaped; widespread and weedy, growing well in poor soils. LEAVES: Deciduous, alternate, pinnately compound; whole leaf 8–36 inches (2–10 dm) long, with many leaflets; leaflets broad, elliptic to lanceolate, 2–8 inches (5–20 cm) long; margins entire, or with few teeth at base or teeth irregularly scattered along margin. FLOWERS: Spring; yellow-green; tiny; in large panicles; usually dioecious. FRUIT: Reddish brown oblong samara (winged seed), ½–2 inches (1.2–5 cm) long; in large pendant clusters. RANGE: Native to China, escaping from cultivation.

338. Tree Tobacco, *Nicotiana glauca*
Solanaceae, Nightshade Family

Slender shrub or small tree; young trunk and branches green; in sandy or clay soils, usually near water; uncommon. LEAVES: Persistent in mild winters, simple, alternate; broadly lanceolate, egg-shaped, or triangular; blade 1–9 inches (2.5–23 cm) long, 1–4½ inches (2.5–11.5 cm) wide; surface smooth, with white waxy coating; margins entire. FLOWERS: Spring or throughout year; yellow; tubular,

5-lobed; 1–2 inches (2.5–5 cm) long. FRUIT: Capsule ¾ inch (2 cm) long with numerous tiny seeds. NOTE: All parts of the plant are poisonous. RANGE: Native of Argentina, escaping cultivation in South and West Texas.

339. Texas Snowbell, *Styrax texanus*
Styracaceae, Snowball Family

Shrub to 13 feet (4 m) tall; along streams, on cliffs above streams; endangered. LEAVES: Deciduous, simple, alternate; blade broad, egg-shaped, or oval to nearly round, 1–3 inches (2.5–7.5 cm) long; top shiny, smooth; bottom very pale or white, coated with minute silky hairs; margins entire. FLOWERS: Spring; white bells; 5 petals; to ¾ inch (2 cm) long. FRUIT: Round dry capsule to ⅜ inch (1 cm) wide. RANGE: Endemic to Edwards Plateau. RELATED SPECIES: *Styrax platanifolius* endemic to Edwards Plateau, has leaf margins entire, wavy, or somewhat lobed, both top and bottom of leaf are densely hairy or smooth, bottom may be pale but not white; *S. youngiae* in the Davis Mountains is possibly extinct; 2 species in East Texas have elliptic or oval leaves, with margins toothed or entire—*S. americanus* with leaves less than 3 inches (7.5 cm) long, and *S. grandifolius* with leaves 2–6 inches (5–15 cm) long. Silverbells (*Halesia* species), in East Texas, have toothed leaves, flowers with 4 petals, and dry, elongated, 2- or 4-winged fruits.

340. Tamarisk, Salt Cedar, *Tamarix* **species**
Tamaricaceae, Tamarisk Family

Shrubs or small to large trees; along waterways, in salt flats, and arid brush; widespread and weedy. LEAVES: Deciduous or evergreen; minute scales; in some species, leaves form drooping thread-like strands. FLOWERS: Year-round; pink or white; tiny; in cylindrical racemes 1–3 inches (2.5–8 cm) long. FRUIT: Tiny capsule; seeds with attached hairs. RANGE: Introduced from Africa and the

Middle East, escaping from cultivation. NOTE: Tamarisks have become damaging weeds in the arid Southwest. Although they provide habitat for birds and reduce soil erosion along rivers, they displace native vegetation along streams and can dry up bodies of water. RELATED SPECIES: Several species hybridize and are difficult to distinguish.

341. Bald Cypress, Sabino, *Taxodium distichum*
Taxodiaceae, Taxodium Family

Deciduous conifer to 110 feet (33.6 m) tall, with large trunk; base of tree swollen; roots may emerge from water in cone-shaped knees; bark shredding or scaly; in swamps and along permanent streams; common. LEAVES: Linear flat needles, ¼–1 inch (6–25 mm) long; needles single; golden or rusty brown in fall. FLOWERS: Spring–fall; minute cones in catkins. FRUIT: Round green cone about 1 inch (2.5 cm) in diameter. RANGE: Eastern U.S. NOTE: Lumbering has eliminated many large trees, which can be more than 1,000 years old. RELATED SPECIES: Montezuma bald cypress (*Taxodium mucronatum*), is a large evergreen in the Lower Rio Grande Valley. Also see pines, pages 260–261.

342. Basswood, Linden, *Tilia americana*
  (*T. caroliniana*)
Tiliaceae, Linden Family

Small to large tree to 92 feet (28 m) tall; in moist woods and along streams; uncommon. LEAVES: Deciduous, simple, alternate; blades often lopsided, ovate to heart-shaped, 2–8 inches (5–20 cm) long; tip pointed; bottom surface hairless to densely hairy; margins sharp-toothed. FLOWERS: Spring–summer; white to yellowish; fragrant; cluster of small flowers attached by slender stem to midpoint of leaflike linear or oblong bract. FRUIT: Dry round drupes ¼ inch (8 mm) in diameter, in clusters hanging from bract. RANGE: Eastern U.S.; Canada; Mexico. NOTE: Flowers make a nice tea; leaf buds tasty.

343. Hackberry, Sugarberry, Palo Blanco, *Celtis laevigata*
Ulmaceae, Elm Family

Small to large tree; bark gray, smooth or with prominent warty ridges; in various wooded habitats, often by streams, along fencerows; widespread and abundant, with several varieties. LEAVES: Deciduous, simple alternate; blade highly variable, usually lanceolate, 1–7 inches (2.5–17 cm) long; tip usually long-tapering, often curving; top smooth or rough; margins entire or toothed (teeth numerous or irregularly scattered). FLOWERS: Emerging with leaves; green; inconspicuous. FRUIT: Red-orange to brown drupe ¼–⅜ inch (6–10 mm) in diameter; flesh thin, sweet, edible. RANGE: Southeastern U.S.; Mexico. RELATED SPECIES: This species hybridizes with netleaf hackberry (*C. reticulata*), a small tree common from the Edwards Plateau west and in scattered populations in East Texas; the leaves are usually less than 3 inches (7.5 cm) long, generally broad, more egg-shaped than lanceolate, and sandpapery rough on the top. *Celtis laevigata* also hybridizes with several other less common species.

344. Granjeno, Desert Hackberry, *Celtis pallida*
Ulmaceae, Elm Family

Shrub typically less than 10 feet (3 m) tall, with straight paired spines and slightly zigzagging branches; bark gray; in dry brush; common. LEAVES: Persistent, simple, alternate; blade variable, oblong, elliptic or egg-shaped, size variable, ¼–1¾ inches (7–45 mm) long, occasionally longer; on some shrubs, all leaves minute, less than ½ inch (1.2 cm) long; surface rough; margins entire or coarsely toothed. FLOWERS: Early spring; white; inconspicuous.

FRUIT: Yellow, orange, or red fleshy drupe ¼–½ inch (6–12 mm) in diameter; sweet, edible. RANGE: Texas to Arizona; Mexico.

(photos 345, 346) Elm, *Ulmus* species (see Key)
Ulmaceae, Elm Family

Small to large trees. LEAVES: Deciduous, simple, alternate; egg-shaped, lanceolate, elliptic, or oblong; straight parallel veins conspicuous on bottom surface; margins sharply and usually doubly toothed. FLOWERS: Minute, inconspicuous; pollen allergenic. FRUIT: Flat, oval or oblong samara (winged seed). NOTE: ornamental species in towns may cause misidentification. SIMILAR SPECIES: Water elm (*Planera aquatica*), in East Texas wetlands, has leaves with singly toothed margins and fruit that is ½ inch (1.2 cm) long, leathery, and covered with warty projections.

KEY TO ELMS

1a. Leaves small, usually less than 3 inches (8 cm) long; sometimes with corky wings on twigs .... go to 3

1b. Most leaves longer than 2½ inches (6.5 cm); never with corky wings on twigs .... go to 2

2a (1b). Top of leaf smooth or somewhat rough; bottom hairless or softly hairy; samaras fringed with hairs ....
American elm, *Ulmus americana*—Large tree; upright limbs form vaselike profile; in East, North Central, and Central Texas; leaf blades 2–6 inches (5–15 cm) long; base of leaf markedly uneven; flowers and fruit in late winter to early spring; now uncommon because of decimation by Dutch elm disease. RELATED SPECIES: The ornamental Siberian elm (*U. pumila*) is widely planted and escaping in western half of Texas, where no native elms grow; it has smooth, dark green leaves usually less than 3 inches (8 cm) long, symmetrical leaf base and samaras without fringe.

2b. Top of leaf sandpapery rough; bottom softly hairy; samaras not fringed with hairs (but center may be slightly hairy) ....
Slippery elm, *Ulmus rubra*—Large tree; in Northeast, North Central, and Central Texas; leaf blade 3–9 inches (8–22.5 cm)

long; base of leaf markedly uneven; flowers and fruit in late winter to early spring; tea of inner bark is an herbal remedy for coughs.

3a (1a). In woods, often in dry limestone soil; from Northeast and North Central, to Central and South Texas, east to Houston; bark often flaky; wings on twigs occasional; leaf stiff, top rough like sandpaper; tip blunt or pointed but not long-tapering; flowers and fruit appearing summer–fall; small to large tree; abundant; samara fringed with hairs . . . . cedar elm, olmo, *Ulmus crassifolia* (photo 346)

3b. In woods, often in sandy soil near water; in East Texas to Tarrant and Bexar counties, with isolated population in Erath County; twigs usually winged; top of leaf smooth to rough; tip pointed to long-tapering; flowers and fruit appearing in the late winter–early spring; small to large tree; common; samaras fringed with hairs . . . . winged elm, *Ulmus alata* (photo 345)

347. Bee Brush, Quebradora, Oreganillo,
Aloysia gratissima
Verbenaceae, Verbena Family

    Small fragrant shrub with straight delicate paired branches (somewhat spine-tipped); in various arid habitats; widespread and common. LEAVES: Deciduous, simple, opposite; blade narrow, elliptic to oblong, ¼–1 inch (7–25 m) long; top rough; bottom pale, hairy; margins entire. FLOWERS: Spring–fall, after rains; white to violet-tinged; minute; 4-lobed; in spikes 1–3 inches (2.5–8 cm) long. RANGE: Texas to California; Mexico, South America. NOTE: An excellent honey plant. The aromatic leaves and flowers have been used as teas. RELATED SPECIES: Variety *schulzae* often has leaves with a few teeth and the bottom rough. Two species have egg-shaped to nearly round leaves, teeth on the margins, and white woolly hairs on the bottom—*A. macrostachya*, in South Texas, has leaves to 1½ inches (4 cm) long and pink, red, or lavender flowers in spikes to 6 inches (1.5 dm) long; *A. wrightii*, in the Trans-Pecos and along the Rio Grande to South Texas, has leaves to ¾ inch (2 cm) long, and white flowers in spikes to 1½ inches (4 cm) long.

348. American Beauty Berry,
French Mulberry, *Callicarpa americana*
Verbenaceae, Verbena Family

Shrub about 3–15 feet (1–4.6 m) tall; in sandy or rocky soil, in woods, often near water; fairly common. LEAVES: Deciduous, simple, opposite or whorled; blade elliptic, oval, or egg-shaped, 3–9 inches (7.5–23 cm) long; aromatic; often hairy; margins toothed. FLOWERS: Spring; white, pink, or lavender; fragrant; minute; in small clusters. FRUIT: Fleshy purple drupes (a white fruiting variety is rare), ¼ inch (7 mm) in diameter; in dense clusters along branches; not edible. RANGE: Southeastern U.S. NOTE: Often used in landscaping.

349. Texas Lantana, Hierba de Cristo,
Lantana urticoides (L. horrida)
Verbenaceae, Verbena Family

Small shrub, may have tiny spines on branches; in various habitats, but most common in sandy soils near coast. LEAVES: Deciduous, simple, opposite or in whorls; blade lanceolate, egg-shaped to nearly round; usually 1–2¼ inches (2.5–6 cm) long, to 1½ inches (4 cm) wide; surface rough, wrinkled; strongly aromatic; margins coarsely toothed. FLOWERS: Spring–fall; yellow, orange, and/or red; tiny; 4- or 5-lobed; in hemispheric multicolored heads 1-3 inches (2.5–7.5 cm) wide. FRUIT: Tight clusters of small fleshy drupes, blue-black when mature; deadly poisonous. NOTE: A popular drought-tolerant ornamental. RANGE: Southwestern U.S.; Mexico. RELATED SPECIES: *Lantana camara*, a native of the West Indies and Florida that escapes cultivation, has leaves to 4 inches (1 dm) long and to 2¾ inches (7 cm) wide and cream, yellow, orange, red, or pink flowers. Other native species, with restricted ranges, have white, pink, or purple flowers.

350. Guayacán, Soapbush, *Guajacum angustifolium*
(Porlieria angustifolia)
Zygophyllaceae, Caltrop Family

Small shrub or small tree about 3–20 feet (1–6 m) tall, with stiff branches; in sun or partial shade, in arid brush, deserts, mountains; common. LEAVES: Evergreen, opposite or clustered, pinnately compound, with numerous leaflets; leaflets linear, oblong or widened at tip, ¼–¾ inch (6–20 mm) long; leaves fold at night and in heat of day; margins entire. FLOWERS: Early spring–summer; purple; fragrant; ½–¾ inch (1.2–2 cm) wide, with 5 petals. FRUIT: Small flattened capsule with 2–4 wings; scarlet flesh covers seed. RANGE: Texas; Mexico. NOTE: The root is used to make soap in Mexico.

351. Creosote Bush, Gobernadora,
Larrea tridentata
Zygophyllaceae, Caltrop Family

Aromatic shrub with numerous slender branches rising from base, about 3–10 feet (1–3 m) tall; abundant in desert and dry plains; often more or less evenly spaced across landscape. LEAVES: Evergreen, opposite, compound, with 2 leaflets joined at base like wings; leaflets to ⅜ inch (1 cm) long; shiny, resinous, smelling like creosote; margins entire. FLOWERS: Almost year-round; yellow; about 1 inch (2.5 cm) in diameter; 5 petals. FRUIT: Tiny round capsule, hairy, silvery to brown, splitting into 5 flat segments. RANGE: Texas to California; Mexico. NOTE: The wax on the leaves contains an antioxidant with potential industrial and pharmaceutical value. ❧

9

Cacti, Agaves, Yuccas, and Other Succulents

KEY TO CACTI, AGAVES, YUCCAS, AND OTHER SUCCULENTS

For woody plants with spines, see Key to Trees and Shrubs, p. 144

1a. Spines, when present, are typically in clusters, emerging from tiny pits or protrusions (areoles, not always visible) on the fleshy surface; stems or pads succulent, fleshy or swollen, with a thick, waxy or horny skin; in most species, flowers showy, colorful; leaves absent in most species .... cactus family (Cactaceae) p. 282–288

1b. Spines, when present, not in clusters; plants with swordlike or dagger-like leaves, or leaves narrowly linear or grasslike .... go to 2

2a (1b). No spines on margins of leaves .... go to 5

2b. Spines on margins of leaves .... go to 3

3a (2b). Leaves waxy, rigid, usually less than 1 inch (2.5 cm) wide; leaf bottom white .... *Hechtia*, p. 298

3b. Not as above .... go to 4

4a (3b). Leaves flexible, linear, typically less than 1 inch (2.5 cm) wide .....
.... sotol (*Dasylirion*), p. 291

4b. Not as above; leaves rigid .... *Agave*, p. 288-290

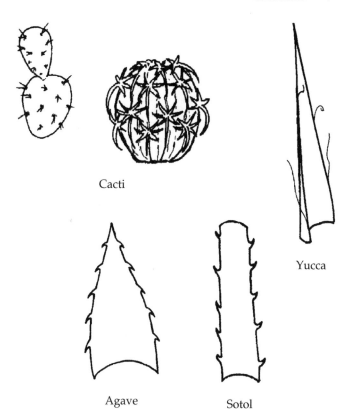

Cacti

Yucca

Agave   Sotol

CACTACEAE, CACTUS FAMILY

352. Texas Hedgehog,
Sclerocactus uncinatus var. *wrightii*
(*Ancistrocactus uncinatus* var. *wrightii*)
Cactaceae, Cactus Family

Barrel cactus to 6 inches (1.5 dm) tall, 4 inches (1 dm) wide; 9–13 ribs; heavy spines in clusters along ribs, central spine to 4 inches (1 dm) or longer, with fishhook tip; in limestone or alluvial soil, in desert; rare, seldom seen east of Pecos River. FLOWERS: Spring–summer; maroon to rusty brown; about 1–1½ inches (2.5–4 cm) long and wide. FRUIT: Oblong or egg-shaped, red, fleshy, to 1 inch (2.5 cm) long. RANGE: Texas, New Mexico; Mexico. RELATED SPECIES: Several other species have fishhook spines. One endangered endemic species.

353. Living Rock, Sunami, *Ariocarpus fissuratus*
Cactaceae, Cactus Family

Unusual spineless cactus, level with or barely above ground, seldom exceeding 4 inches (1 dm) in diameter; wrinkled, triangular protrusions radiate around center; protected by camouflage, low growth habit, and horny epidermis; on loose rocky limestone slopes in arid sunny desert; does not tolerate much moisture or shade; fairly common in Big Bend, rare elsewhere and often overlooked. FLOWERS: Late summer–early fall; white, pink, or red, emerging from woolly center; 1–2 inches (2.5–5 cm) wide. FRUIT: Tiny, greenish or white, turning brown and dry and disintegrating when mature. RANGE: Texas; Mexico.

354. Horse Crippler, Devil's Head,
Manco Caballo, *Echinocactus texensis*
Cactaceae, Cactus Family

Barrel cactus 2–8 inches (5–20 cm) tall, 4–12 inches (1–3 dm) wide, broad and flattened or hemispheric; 13–27 narrow, often irregular ribs; rigid arching spines clustered on ribs; in dry soil; widespread but not common. FLOWERS: Spring–summer; pink, lavender, or reddish orange (sometimes white); 1–2¼ inches (2.5–6 cm) in diameter; emerging from woolly tip of plant. FRUIT: Round or oval, red, fleshy, ¾–2 inches (2–5 cm) long. RANGE: Texas, Oklahoma, New Mexico; Mexico. RELATED SPECIES: *Echinocactus horizonthalonius*, in the Chihuahuan Desert, is rounded to pyramidal or cylindrical and usually has 8 broad, rounded ribs (5–13 possible) covered with heavy interlocking spines; one variety endangered.

355. Strawberry Cactus, Pitaya,
Echinocereus stramineus
Cactaceae, Cactus Family

Cylindrical cactus with stems to 1 foot (3 dm) long, 2–3 inches (5–7.5 cm) wide; ribs mainly 10–12, covered with dense clusters of spines ⅜–3½ inches (1–9 cm) long; many stems (up to several hundred) in a hemispheric mound 2–4 feet (6–12 dm) across, in dry limestone soil, gravel, or sand, in grasslands, deserts, mountains; fairly common. FLOWERS: Spring–summer; spectacular displays of pink to reddish purple flowers, many blooming at once; 3–5 inches (8–12.5 cm) long; emerging from sides of stem. FRUIT: Round, red or purple, fleshy, to 2 inches (5 cm) wide; spines falling off when mature; sweet and edible. RANGE: Trans-Pecos, Texas; New Mexico; Mexico. RELATED SPECIES: Of 5 endemic species most are threatened or endangered.

356. Rainbow Cactus,
Echinocereus dasyacanthus
(Echinocereus pectinatus var. *neomexicanus)*
Cactaceae, Cactus Family

Cylindrical cactus with single or few stems, to 14 inches (3.5 dm) tall, 2–4 inches (5–10 cm) wide; 12–22 narrow ribs with many clusters of tiny interlocking spines hiding surface; spines white, yellowish, tan, pink, or reddish brown, with reddish or brown tips, often giving cactus banded or rainbow appearance; on dry limestone soil, in grasslands and Chihuahuan Desert; fairly common. FLOWERS: Spring–summer; brilliant yellow, pink, violet, or redpurple; 2–5½ inches (5–14 cm) long and wide; emerging from ribs on side of stem. FRUIT: Round, reddish brown to purple, 1–2 inches (2.5–5 cm) wide; spines falling when mature. RANGE: Texas, New Mexico; Mexico. RELATED SPECIES: Several other varieties in Texas.

357. Lace Cactus,
Echinocereus reichenbachii var. *reichenbachii*
(E. caespitosus)
Cactaceae, Cactus Family

Small cylindrical cactus 2–8 inches (5–20 cm) tall, about 2 inches (5 cm) wide, often branching and clump-forming; 10–19 ribs covered with many clusters of comblike spines, the longest about ¼ inch (6 mm) long, may be interlocking; on sunny slopes, usually in limestone soil but also in sandy granite soil of Llano area; several varieties, some common. FLOWERS: Spring; purple or deep pink; 2–4 inches (5–10 cm) wide; flowers emerging from ribs on side of stem. FRUIT: Round or egg-shaped, green, covered with spines. RANGE: Southwestern U.S.; Mexico. RELATED SPECIES: 13 other varieties in Texas; one endangered.

358. Claret Cup Cactus, *Echinocereus triglochidiatus*
Cactaceae, Cactus Family

Clump-forming cactus with few to numerous stems 2–16 inches (5–40 dm) long; 5–12 ribs lined with clusters of few to numerous spines; in sun or partial shade, on dry hills, ledges, mountains; fairly common, with numerous varieties. FLOWERS: Early spring–summer; brilliant to pale red or orange-red; 1–2½ inches (2.5–6.5 cm) wide; petals fleshy, waxy; flowers emerging from sides of stem; remaining open for several days. FRUIT: Green or red, to 1½ inches (4 cm) in diameter; some varieties lose spines when mature. RANGE: Southwestern U.S.

359. Nipple Cactus,
   Biznaga de Chilitos, *Mammillaria heyderi*
   Cactaceae, Cactus Family

Round flattened cactus to 6 inches (1.5 dm) wide or wider, protruding about 2 inches (5 cm) above ground; covered with rows of nipplelike projections tipped with clusters of straight spines; spines numerous but not obscuring nipples; sap milky; usually in partial shade, in rocky soil, in deserts and grasslands, along sandy beaches; uncommon, with several varieties. FLOWERS: Early spring; white or pink; about 1 inch (2.5 cm) wide; emerging from axils of nipples; may form ring around the plant. FRUIT: Red, fleshy, to 1½ inches (4 cm) long. RANGE: Southwestern U.S. RELATED SPECIES: Above characters distinguish this from other species.

360. Cane Cholla, Cane Cactus, *Opuntia imbricata*
   Cactaceae, Cactus Family

Shrubby or treelike cactus 3–10 feet (1–3 m) or taller; jointed cylindrical branches, about 1–1½ inches (2.5–4 cm) wide near tips, with knobby projections giving branches appearance of coarsely braided rope; old stems with woody skeleton; numerous spines per cluster; spines covered with papery sheath; glochids (tiny hairlike spines) inconspicuous or absent; on dry rocky or sandy soil, in grasslands, deserts; locally abundant, with 2 varieties. FLOWERS: Spring-summer; purple to rose; on branch tips. FRUIT: Yellow when mature, about 1–1¾ inches (2.5–4.5 cm) long, deeply ridged, with glochids. RANGE: Southwestern U.S.; Mexico. RELATED SPECIES: Above characters distinguish it from other *Opuntia* species.

361. Tasajillo, Pencil Cactus, *Opuntia leptocaulis*
Cactaceae, Cactus Family

Shrubby cactus 2–5 feet (6–15 dm) or taller; jointed stems pencil thin, ⅛–¼ inch (3–6 mm) wide near tips, about 1½ inches (4 cm) near base; joints detach easily; 1 or no spines (sometimes 3) per areole, covered with papery sheath; glochids (tiny hairlike spines) few, inconspicuous; in dry soil; common and weedy. FLOWERS: Spring–summer; greenish yellow; less than 1 inch (2.5 cm) wide. FRUIT: Bright red, orange-red, or yellow; fleshy; ½–1 inch (1.2–2.5 cm) long; with glochids; surface not ridged. RANGE: Southwestern U.S.; Mexico. RELATED SPECIES: *Opuntia kleiniae*, in the Trans-Pecos, has joints ⁵⁄₁₆–½ inch (8-12 mm) wide near tips, greenish purple to pink flowers, and fruits with or without ridges.

362. Texas Prickly Pear,
Opuntia engelmannii (O. lindheimeri)
Cactaceae, Cactus Family

Large, shrubby cactus 2–10 feet (6–30 dm) tall or taller, with flat jointed stems forming pads 8–14 inches (2–3.5 dm) in diameter; pads circular or broadly egg-shaped (cow-tongue prickly pear, *O. lindheimeri* var. *linguiformis*, has pads greatly elongated); spines mostly yellow, usually 1–8 per areole, ½–3 inches (1.2–7.5 cm) long (some varieties spineless); glochids (tiny barbed hairlike spines) in clusters along edge of pad and usually at base of long spines; abundant, densely populating overgrazed rangeland; several varieties. FLOWERS: Spring–summer; yellow, yellow-orange, or red; showy; large. FRUIT: Pear- or egg-shaped (widest at top), red or purple, fleshy, 1–3 inches (2.5–7.5 cm) long; with glochids. RANGE: Western U.S.; Mexico. RELATED SPECIES: Numerous other less common species. *Opuntia phaeacantha* in West Texas, is very similar. NOTE: Tiny succulent conical leaves form in early spring, then soon fall off. The glochids can cause skin rash or more severe allergic reactions. The edible sweet fruit is good for jelly; the young pads, also edible, are best when boiled, then sautéed in butter. One endemic variety and 2 endemic *Opuntia* species may be endangered.

363. Dog Cholla, Clavellina, *Opuntia schottii*
    Cactaceae, Cactus Family

Low cactus under 1 foot (3 dm) high; jointed stems in single chains or dense mats 1–2 feet (3–6 dm) across; joints cylindrical or club-shaped, 1–3 inches (2.5–7.5 cm) long, to 1 inch (2.5 cm) wide at tips, with knobby projections giving stems a ropy appearance; spine to 2½ inches (6.2 cm) long, about 14 or fewer per areole; glochids few (bristly spines at base of large spines); on gravelly hills; common, with 2 varieties. FLOWERS: Early spring–summer; yellow; showy; about 1–2½ inches (2.5–6.5 cm) wide. FRUIT: Ropy, yellow when mature, 1–1¾ inches (2.5–4.5 cm) long; with large glochids. RANGE: Texas. RELATED SPECIES: The rare *Opuntia kunzei* from El Paso to Candelaria (Presidio County), has joints 3–8 inches (7.5–20 cm) long; and very spiny fruit.

364. Purple Prickly Pear, *Opuntia macrocentra*
    (O. violacea)
    Cactaceae, Cactus Family

Shrubby cactus 2–3 feet (6–9 dm) tall, with jointed flattened stems; pads 4–8 inches (1–2 dm) wide, green or blue-green to purple (purple increases in winter or dry summers); 1–3 spines per cluster, 1½–7 inches (4–17.5 cm) long, spines often black; glochids (tiny hairlike spines) inconspicuous; in desert soils. FLOWERS: Spring–summer; yellow with red center; often fading to pink; large, showy. FRUIT: Red, fleshy, to 1½ inches (4 cm) long; edible. RANGE: Texas, New Mexico, Arizona; Mexico.

AGAVACEAE, AGAVE FAMILY

(photos 365, 366) Agave, Century Plant, *Agave* species (see Key)
Agavaceae, Agave Family

Succulent shrubs without trunks; composed of basal rosettes of stiff swordlike leaves; in arid soil, in deserts and mountains. LEAVES: Green year-round; spines protrude from tip and, in most species, line margins. FLOWERS: Green, yellow, or red; 6 tepals form tube; in large clusters at top of stalk; stalk 10–30 feet (3–9 m) tall in some species; blooms appear once, at end of life span of 8–30 years. FRUIT: Dry capsule; seeds flat, black. NOTE: The caustic sap in leaves of most species will burn the skin. Some species are valued for food, fiber, medicine, and alcohol production; some species are toxic. Some references classify all plants in the Agavaceae as members of the Liliaceae, Lily Family.

KEY TO *AGAVE* SPECIES

Hybrids may create difficulties in identification.

1a. Found in the Trans-Pecos, east to Val Verde County . . . . go to 5

1b. Found in South Texas, or in cultivation if elsewhere . . . . go to 2

2a (1b). Small agave; leaves linear to lanceolate, 12–28 inches (3–7 dm) long, to 2¼ inches (6 cm) wide; flowers yellow or green, in spikes (inflorescence not branching) . . . .
Agave lophantha—In Starr and Zapata counties, also cultivated; leaves bright light green or yellow-green, often with light green midstripe; margins wavy or scalloped and lined with spines; the plant is toxic.

2b. Medium to large agave; leaves 2–6 feet (6–20 dm) long, considerably broader than 2 inches (5 cm) at base; flowers yellow, in large panicles (branching clusters) . . . . go to 3

3a (2b). Leaf margins smooth, unarmed, at least in upper third; leaves green, 3–5 feet (1–1.6 m) long, 4½–7 inches (1.2–1.8 dm) wide at base; cultivated in Webb County .... *Agave weberi*

3b. Leaf margins heavily armed with spines .... go to 4

4a (3b). Leaf surface generally smooth to the touch ....
Century plant, maguey, *Agave americana*—wild only in far South Texas; cultivated throughout Texas, with numerous varieties; leaves 3–6 feet (1–2 m) long, 6–10 inches (1.5–2.5 dm) wide at base; light green to glaucous gray to variegated with yellow or white; Mexican villagers obtain a beverage, aguamiel, from this agave by removing the leaf bud and allowing a liquid to accumulate; leaves provide fibers for rope, but the caustic pulp must be rotted off in water.

4b. Leaf surface rough to the touch; light green to grayish leaves, 24–44 inches (6–11 dm) long, 5–7 inches (1.2–1.8 dm) wide at base; in Starr, Webb, and Zapata counties .... *Agave parryi* ssp. *parryi* (*Agave scabra*)

5a (1a). Small agave; leaves linear to lanceolate, 8–24 inches (2–6 dm) long, to 1½ inches (4 cm) wide; yellow or reddish flowers in tall spikes; abundant in Chihuahuan Desert, Trans-Pecos and western Edwards Plateau ....
Lechuguilla, *Agave lechuguilla* (photo 366)—Leaves yellow-green, often with reddish or dark green streaks on back. The plant has caused death in livestock. Steroids have been synthesized from the toxic saponins. Stiff leaf fibers are used for brushes and rope.

5b. Small to large agave; leaves wider than 1½ inches (4 cm) .... go to 6

6a (5b). Medium to large agave; leaves broad, 4–12 inches (1–3 dm) wide at base, egg-shaped or triangular with tapering tip, typically 2–4 times longer than wide ....
Agave havardiana (photo 365)—in mountains, particularly abundant from Davis Mountains south; leaves 12–34 inches (3–8.6 dm) long, glaucous gray or light green or yellowish, teeth pointing downward; flowers in spring to summer, yellow, in broad panicles (branching clusters).

6b. Small to medium agaves; leaves usually under 5 inches (1.2 dm) wide at base; typically 3–9 times longer than broad .... go to 7

7a (6b). Flowers in raceme or spike (cluster not branching) ....
Agave glomeruliflora (*A. chisosensis*)—In mountains, Brewster and Culberson counties; rare and endangered; leaves variable, (linear, lanceolate, or triangular), 14–22 inches (3.5–5.5 dm) long, 2–4 inches (5–10 cm) wide at base, green or glaucous; flowers yellow.

7b. Flowers in large panicles (branching clusters) .... go to 8

8a (7b). Blooming in fall; panicles narrow (branches bend upward); flowers 1½–2¼ inches (4–5.5 cm) long ....

Mescal, *Agave gracilipes*—In Trans-Pecos mountains; uncommon; leaves lanceolate, 3–6 times longer than wide, 7–15 inches (1.8–3.8 dm) long, 1½–3 inches (4–7.5 cm) wide at base, glaucous gray to yellow-green; flowers yellow or red.

8b. Blooming in spring; panicle branches broad (spreading horizontally); flowers slightly longer, 2¼–2⅝ inches (5.5–6.7 cm) ....
Mescal, *Agave parryi* ssp. *neomexicana*—In northern Trans-Pecos mountains; common; leaves lanceolate, 3–5 times longer than wide, 6–18 inches (1.5–4.5 dm) long, 1½–4½ inches (4–12 cm) wide at base, glaucous gray to light green; flowers yellow, buds red or orange. Mescalero Apaches ate the cooked heart of mescal.

367. Sotol, *Dasylirion* **Species**
Agavaceae, Agave Family

Shrubs composed of dense basal rosette of long swordlike leaves; usually trunkless, though some may have short trunk; in deserts and mountains, on dry, rocky slopes; locally abundant. LEAVES: Green year-round; linear, about 2–4 feet (6–12 dm) long; most of leaf under 1 inch (2.5 cm) wide; margins lined with hooked or straight spines, but no spine at tip. FLOWERS: Spring–summer, flowering every few years; whitish or greenish; tiny; in dense clusters on stalk 10–16 feet (3–5 m) tall; dioecious. FRUIT: Tiny, 3-winged, leathery; enclosing 1 seed. RANGE: Southwestern U.S.; Mexico. NOTE: The leaves are a source of fiber. RELATED SPECIES: The 4 Texas species are difficult to distinguish; all grow in the Trans-Pecos. *Dasylirion texanum* also occurs on the Edwards Plateau; *D. leiophyllum* is the most common Trans-Pecos species; *D. heteracanthium*, in the southern Trans-Pecos, may be a hybrid of these two. *Dasylirion wheeleri* is found in El Paso, Culberson, and Hudspeth counties.

368. Red Yucca, Samandoque, *Hesperaloë parviflora*
Agavaceae, Agave Family

Low rosette of fleshy swordlike leaves; in dry brush; uncommon. LEAVES: Green year-round; swords linear, 2–5 feet (6–15 dm) long, about 1 inch (2.5 cm) wide; margins entire, with loose fibers, edges rolling inward; tip may or may not be spiny. FLOWERS: Spring–fall; pink to red; 6 tepals form tubular or bell-shaped flower; in clusters along stalk 4–8 feet (1.2–2.5 m) tall. FRUIT: Small capsule; seeds flat, black. RANGE: Texas; Mexico. NOTE: Red yucca is often used in landscaping. The red flowers distinguish it from the true yuccas. Leaves useful in basket making.

369. Beargrass, Sacahuista, *Nolina* species (see Key)
Agavaceae, Agave Family

Shrub consisting of a dense cluster of long, grasslike leaves; occasionally with a short trunk. LEAVES: Green year-round, narrowly linear, 2–8 feet (6–24 dm) long; erect, or drooping and grasslike, less than 1 inch (2.5 cm) wide. FLOWERS: Spring, summer; white or tinged with purple; tiny, with 6 tepals; in panicles (branching clusters) emerging from center of plant. FRUIT: Tiny capsule with brown or black seed. NOTE: Toxic to livestock. Leaves good for baskets.

KEY TO *NOLINA* SPECIES

1a. Leaf margins with fine sawlike teeth . . . . go to 3

1b. Leaf margins smooth or with few teeth . . . . go to 2

2a (1b). Found east of the Trans-Pecos . . . .
Nolina texana—In upper Rio Grande Plains, Edwards Plateau, southern Panhandle, and Trans-Pecos; widespread and common. LEAVES: To about 4 feet (1.2 m) long, to ³⁄₁₆ inch (5 mm) wide; drooping and grasslike. FLOWERS: White, occasionally purple-tinged; flowering stalk 1–2 feet (3–6 dm) tall, stiff, held within or barely protruding above foliage.

2b. Found in the Trans-Pecos . . . . one of the following:
(i) *Nolina arenicola*—In deep sand, endemic to Culberson and Hudspeth counties; rare and endangered. LEAVES: Flat or rounded, to about 4¼ feet (1.3 m) long, about ³⁄₁₆ inch (5 mm) wide; grasslike; drooping or erect. FLOWERS: White, occasionally purple-tinged; flowering stalk to 5 feet (1.5 m) tall, stiff.

(ii) *Nolina micrantha*—In sandy or rocky soil, in mountains of Culberson and Hudspeth counties; common. LEAVES: To about 4¼ feet (1.3 m) long, about ³⁄₁₆ inch (5 mm) wide; grasslike; drooping or erect. FLOWERS: Usually purple-tinged; flowering stalk to 3 feet (1 m) tall, flexible, usually enclosed within foliage.

(iii) *Nolina texana* (see 2a.)

3a (1a). Endemic to Edwards Plateau; leaves to 3 feet (1 m) long, ⅙–½ inch (4–12 mm) wide; drooping or erect; flowering stalk 2–6 feet (6–20 dm) tall . . . . *Nolina lindheimeriana*

3b. Found in the Trans-Pecos . . . . go to 4

4a (3b). Found throughout the Trans-Pecos; leaves 3–8 feet (1–2.5 m) long, ⅜–1 inch (1–2.5 cm) wide; typically erect; flowering stalk 2–8 feet (6–25 dm) tall, partially or entirely enclosed within foliage . . . .
. . . . *Nolina erumpens* (photo 369)

4b. Rare, restricted to northwestern Trans-Pecos; leaves less than 4 feet (1.2 m) long, and less than ½ inch (1.2 cm) wide; drooping or erect; flowering stalk to about 6 feet (1.8 m) tall . . . . *Nolina microcarpa*

(photos 370 through 373) Yucca, Spanish Dagger, Izote, Soyate, *Yucca* species (see Key)
Agavaceae, Agave Family

Drought-tolerant succulents with swordlike leaves; abundant in Chihuahuan Desert, but species occur all over Texas. Two basic forms in Texas—tree yuccas, with trunklike stems, and low-growing trunkless shrubs (young tree yuccas may be confused for trunkless species; some species that are usually trunkless can develop short trunk.) LEAVES: Green year-round; most longer than 1 foot (3 dm); tipped with needle-sharp spines; no spines on margins. FLOWERS: Flowering yearly (agaves bloom only once); white, cream, or greenish, some tinged with reddish-purple; 6 tepals, 1–4 inches (3–10 cm) long, depending on species; tall flower stalk yields large panicle (branching cluster); in a few species, flower stalk is a raceme (not branching). FRUIT: Capsule, either dry and woody (usually erect) or fleshy and leathery (pendant); 1–4 inches (2.5–10 cm) long; sweet fruit of *Yucca baccata* to 9 inches (23 cm) long; seeds usually black. NOTE: Unregulated collecting has resulted in the decimation of many populations of wild plants. Yuccas have long been valued by Native Americans for the fiber from the leaves, the soap from the roots, and the edible fruits of the fleshy-fruited species; the white flowers also are edible. Warning: Similar looking agaves have caustic leaves.

KEY TO *YUCCA* SPECIES

Several species hybridize. Native and non-native species are commonly used as ornamentals outside their natural ranges. Both factors add to the difficulty in distinguishing species. The key does not include non-native ornamentals.

1a. Leaf margins with fine sawlike teeth (or occasionally smooth) and no loose threads; margins often yellow or reddish brown, horny . . . . go to 2

1b. Leaf margins smooth or slightly rough, but without sawlike teeth; usually with obvious loose threads (sometimes threads wear off on older leaves); margins green, white, or sometimes brown, but not yellow . . . . go to 5

2a (1a). Leaves under ¾ inch (2 cm) wide, flexible; plant with or without trunk . . . . go to 4

2b. Leaves to 2 inches (5 cm) wide, flexible to somewhat stiff but still bendable; plant trunkless . . . . go to 3

3a (2b). Leaves noticeably twisted or wavy, dark green or yellow-green . . . Twisted-leaf yucca, *Yucca rupicola* (photo 372)—Common in and endemic to the Edwards Plateau, north to Bell County, in rocky limestone soil; leaves 1–2½ feet (3–7.5 dm) long, ¾–2 inches (2–5 cm) wide; margins saw-toothed; flowers in spring; panicle held well above foliage; fruit a dry woody capsule.

3b. Leaves straight (new growth may be twisted); coated with white waxy coating (a bloom) that makes foliage appear gray-green or bluish . . . Pale-leaf yucca, *Yucca pallida* (photo 371)—endemic to North Central Texas, from Wise County and Dallas to Hays County, in rock outcrops or rocky prairie; leaves 6–24 inches (1.5–6 dm) long, ⅜–2 inches (1–5 cm) wide, margins saw-toothed or smooth; flowers in spring; panicle held well above foliage; fruit a dry woody capsule.

4a (2a). Plant with trunks to 15 feet (4.5 m) tall, with a variety near Langtry; trunkless to 3 feet tall . . . . Beaked yucca, Thompson yucca, *Yucca thompsoniana* (*Y. rostrata*)—In western Edwards Plateau to Pecos and Brewster counties; on rocky hills, in canyons, on mountain slopes; common but populations scattered; leaves 8–24 inches (2–6 dm) long, margin saw-toothed or smooth; flowers in spring; bottom of large panicle beginning within or above foliage; fruit a dry woody capsule.

4b. Plant trunkless . . . .

San Angelo yucca, *Yucca reverchonii*—Endemic to western Edwards Plateau; uncommon; leaves 10–20 inches (2.5–5 dm) long, straight, surface rounded, yellow-green with a slight bloom, margins saw-toothed; flowers in spring; panicle held above foliage; fruit a dry woody capsule.

5a (1b). Leaves easily bent, usually less than 1½ inches (4 cm) wide . . . go to 8

5b. Leaves rigid, 1–3½ inches (2.5–9 cm) wide . . . . go to 6

6a (5b). Plant usually trunkless (occasionally with short trunk) . . . .

Banana yucca, datil, *Yucca baccata*—From northern Trans-Pecos to western Edwards Plateau; in mountains, grasslands, juniper/oak woodlands; leaves 1–3½ feet (3–10.5 dm) long, to 2½ inches (6.5 cm) wide, margins with tough curling threads; all or part of flower panicle held within foliage; fruit fleshy, banana-like, 3–9 inches (7.5–23 cm) long, sweet.

6b. Plant usually with treelike trunk . . . . go to 7

7a (6b). Very large yucca, 5–25 feet (1.5–7.6 m) tall; only in a few scattered populations in the Trans-Pecos; abundant at Dagger Flats in Big Bend . . .

Giant yucca, giant dagger, *Yucca faxoniana* (including *Y. canerosana*)—leaves 2–4 feet (6–12 dm) long, to 3 inches wide (7.5 cm); often forming large symmetrical crown; flowers in spring to fall, large panicle partially within or held slightly above foliage; fruit fleshy or leathery, sweet.

7b. Large yuccas, 3–15 feet (1–4.5 m) tall or taller; common or abundant, east or west of the Pecos . . . . . either of these 2 intergrading species:

(i) Torrey yucca, *Yucca torreyi*—Abundant throughout the Trans-Pecos, extending onto Edwards Plateau and upper Rio Grande Plains; occasionally trunkless; plants often solitary; leaves 2–4½ feet (6–13 dm) long, usually under 2¼ inches (6 cm) wide, margins with tough threads; on many shrubs, majority of green leaves erect, dead leaves hang at angles, giving shrub an untidy appearance; flowers in spring; panicle partially to entirely enclosed within foliage; fruit fleshy or leathery.

(ii) Trecul yucca, Spanish dagger, palma pita, *Yucca treculeana* (photo 373)—From southern Gulf Coast, through South Texas and southern Edwards Plateau; often clump-forming; leaves 2–4½ feet (6–13 dm) long, 1½–3½ inches (4–9 cm) wide; leaf margins smooth or slightly rough, with or without loose threads; leaves may form a symmetrical crown; flowers in late winter to spring; panicle partially to entirely enclosed within foliage; fruit fleshy or leathery.

8a (5a). Plants with treelike trunks 3–15 feet (1–4.5 m) tall or taller (some plants may be trunkless) . . . .

Soaptree yucca, palmella, *yucca elata* (photo 370)—In desert in. Trans-Pecos; common and widespread; leaves 1–3 feet (3–9 dm) long, usually ⅛–⅝ inch (4–15 mm) wide, occasionally to 1 inch (2.5 cm) wide; leaf margins white; flowers in late spring to summer; flower stalk to 15 feet (5 m) tall; flowers held well above foliage; fruit a dry woody capsule.

8b. Plants trunkless or with short trunks, usually under 3 feet (1 m) tall . . . . go to 9

9a (8b). Flower cluster held well above foliage (distance between flowers and foliage greater than length of flower cluster) . . . . go to 13

9b. Flower cluster within or held slightly above foliage (distance between flowers and foliage less than length of flower cluster) . . . go to 10

10a (9b). Found in the Trans-Pecos or north to the Panhandle; leaves typically less than ½ inch (1.2 cm) wide . . . . go to 12

10b. Found in Central or North Central Texas; leaves typically ¼–1 inch (6–25 mm) wide . . . . go to 11

11a (10b). Endemic to Somervell County (in North Central Texas), in sandy soil . . . .

Yucca necopina—Endangered; flowers in racemes or narrow panicles, stalk 3–9 feet (1–3 m) tall; fruit a dry woody capsule.

11b. Widespread, from Central and North Central Texas east onto Blackland Prairie, in rocky or sandy soil, in prairies and plains . . . .

Arkansas yucca, *Yucca arkansana*—Usually trunkless; leaves 6–30 inches (1.5–7.5 dm) long, ¼–1½ inch (6–40 mm) wide, margins white; flowers in spring, in racemes or with a few spreading branches near base, stalk less than 5 feet (1.6 m) tall; fruit a dry woody capsule.

12a (10a). Found throughout Panhandle . . . .

Small soapweed yucca, *Yucca glauca* var. *glauca* (*Yucca angustifolia*)—Trunkless or with short trunk to 1 foot (3 dm); often in sand; leaves 8–36 inches (2–10 dm) long, ³⁄₁₆–½ inch (5–12 mm) wide, margins white; flowers in late spring, in a panicle but often not branching at top; fruit a dry woody capsule.

12b. Endemic to the Panhandle and northern Trans-Pecos . . . .

Plains yucca, *Yucca campestris*—Trunkless or with stem to 3 feet (1 m); usually in deep sand and dunes; leaves to 2 feet (6 dm) or occasionally 3 feet (1 m) long, ⅛–⅜ inch (4–10 mm) wide, wiry, margins white; flowers in spring; fruit a dry woody capsule. May be a hybrid of *Y. elata* and *Y. constricta*.

13a (9a). Found in East Texas, east of Blackland Prairie; in sandy soils . . . . .
 Louisiana yucca, Gulf Coast Yucca, *Yucca louisianensis* (*Y. free-manii*)—Throughout East Texas, common in Northeast Texas, in dry, sandy or sandy-clay soil; usually trunkless but may have short trunk; leaves 1–3½ feet (3–10 dm) long, ¼–1 inch (7–25 mm) or occasionally to 1⅝ inches (4.2 cm) wide; leaf margins white or green; flowers in late spring to summer; branches of panicle hairy or not (hand lens); fruit a dry woody capsule.

13b. Found along southern coast or in South or Central Texas, to southeast edge of Trans-Pecos . . . . one of the following:

 (i) Buckley yucca, *Yucca constricta*—Common, in limestone outcrops or rocky prairie, trunkless or with short trunk or prostrate stem, trunk rarely to 6 feet (2 m); leaves flexible, 8–32 inches (2–8 dm) long, ⅛–¾ inch (4–20 mm) wide, margins white or green; flowers in spring; style green; panicle branches hairless (hand lens); fruit a dry woody capsule.

 (ii) *Yucca tenuistyla*—Endemic from central coast to southern Edwards Plateau; trunkless; style white; panicle branches hairy or not; possibly a variety of *Y. constricta*.

BROMELIACEAE, PINEAPPLE FAMILY

374. Hechtia, False agave,
 Hechtia texensis (*H. scariosa*)
 Bromeliaceae, Pineapple Family

 Low succulent shrub composed of rosette of spiny daggerlike leaves; often forms large circular colonies; on limestone desert hills. LEAVES: Green year-round; stiff waxy daggers; typically less than 16 inches (4 dm) long and 1 inch (2.5 cm) wide near base; often reddish-tinged, with white scales on bottom surface; margins lined with curved spines. FLOWERS: Late winter–spring; tiny; 3 white petals; sepals white or papery; clustered on branches of stalk 2–6 feet (6–20 dm) tall. FRUIT: Small woody capsule. RANGE: Texas; Mexico. RELATED SPECIES: Guapilla (*Hechtia glomerata*), in Starr and Zapata counties. ❧

10

Miscellaneous Plants, Weeds, and Growths on Trees

WEEDS

---

375. Giant Ragweed, *Ambrosia trifida*
Asteraceae, Sunflower Family

Annual weed 3–16 feet (1–5 m) or taller by late summer; sap red; along streams, in disturbed areas; abundant. LEAVES: Simple, opposite (some alternate); blade 2–8 inches (5–20 cm) long; surface sandpapery rough; lower leaves usually with 3-5 fingerlike lobes; upper leaves smaller, unlobed; aromatic; margins toothed. FLOWERS: Late summer–fall; inconspicuous, lacking petals; in tiny green cups arranged on tall spikes; monoecious. RANGE: Most of U.S., Mexico, Canada. RELATED SPECIES: 7 other species; most less than 5 feet (1.8 m) tall, with leaves very different from those of giant ragweed. NOTE: Ragweeds are wind-pollinated, and the abundant pollen causes fall hay fever and asthma. An infusion of the leaves is useful for treating poison ivy.

376. Cocklebur, Abrojo, *Xanthium strumarium*
Asteraceae, Sunflower Family

Bushy annual weed 1–6 feet (3–20 dm) tall; stem and leaves usually rough; in moist soil, disturbed ground along waterways; abundant. LEAVES: Simple, mostly alternate; blade large, may be as broad as long, on long petiole; margins toothed or lobed. FLOWERS: Summer–fall; inconspicuous round green heads; burlike. FRUIT: Brown oval or oblong bur about 1 inch (2.5 cm) long, covered with hooked barbs, clinging to passing mammals. RANGE: Throughout Texas and much of world. RELATED SPECIES: In western half of Texas, intergrades with *Xanthium spinosum*, which has a spine at base of leaf. NOTE: Foliage and fruit have caused poisoning and mechanical damage to livestock.

377. Carrizo, Giant Reed, Georgia Cane, *Arundo donax*
Poaceae, Grass Family

Colony-forming cane 6–20 feet (2–6 m) tall; canes 1–2 inches (2.5–5 cm) thick; in sandy soil and disturbed ground, along rivers and lakeshores; locally abundant. LEAVES: Blade grasslike, lanceolate or linear, 1–2 feet (3–6 dm) long; sheathing the cane. FLOWERS: Spring–fall; minute; in large white clusters at tips of canes. RANGE: Probably from Mediterranean, now naturalized in southern U.S. NOTE: The canes provide the reeds for woodwind instruments. RELATED SPECIES: Giant cane (*Arundinaria gigantea*), an uncommon East Texas native, has much smaller leaves. Common reed (*Phragmites communis*), a smaller cane, grows worldwide and is used in papermaking.

378. Cattail, *Typha* species
Typhaceae, Cattail Family

Colony-forming aquatic perennials; erect stems rising from extensive underwater roots; in shallow water of ditches, marshes, ponds, stream banks; widespread and weedy. LEAVES: Grasslike, sheathing the stem near base; blade 2–7 feet (6–21 dm) tall, ¼–1 inch (6–25 mm) wide. FLOWERS: Spring; minute; green; in dense spikes, with male spike above female spike. FRUIT: Minute seeds attached to threadlike hairs or floss, forming a brown cylindrical spike. RANGE: 3 species in Texas, with colonies scattered throughout the state. NOTE: The seeds, young shoots, and other parts can be used for food. The leaves and water-resistant floss are good for fiber.

GROWTHS ON TREES

379. Galls

Swollen growths found on many different species of trees, shrubs, or herbaceous plants; sometimes mistaken for seeds or nuts. An insect bores a hole into a plant and deposits her eggs inside. The plant swells up around the eggs, forming a gall. When the eggs hatch, the larvae feed on the inside of the gall. Eventually, the adult insect bores an escape hole and flies away. Some species of insects use only certain species of plants for egg laying. Galls look very different on different types of plants. Oak galls are useful as dyes and mordants.

(photos 380, 381) Spanish Moss, Pastle,
  Tillandsia usneoides
  Bromeliaceae, Pineapple Family

Drooping masses of gray-green threadlike fibrous leaves; often forming dense curtains hanging from tree limbs, also on fences and wires; in humid forests or restricted to waterways in dry regions; abundant in some areas. FLOWERS: Late winter–spring; inconspicuous, minute. RANGE: Southeastern U.S., through South America. RELATED SPECIES: Ball moss (*Tillandsia recurvata*, photo 380), adapted to drier conditions, forms small compact balls on tree limbs, wires, and rocks in the western half of Texas. NOTE: Not parasitic and not really mosses, *Tillandsia* species are epiphytes, extracting water from the air. Fibers used to be valued as stuffings for automobile cushions.

382. Mistletoe, *Phoradendron* species (see key)
Viscaceae, Mistletoe Family

Green or yellow-green branching perennials, to 3 feet (1 m) wide or more; parasitic; roots embedded in tree branches. LEAVES: Evergreen, simple, opposite (reduced to scales in some species). FLOWERS: Greenish, inconspicuous, on short spikes. FRUIT: Tiny white to pinkish translucent berries. NOTE: All parts are deadly poisonous on most species. Though producing their own starches through photosynthesis, mistletoes rely on trees for water and minerals. Dense infestations can weaken the tree, making it more susceptible to insect infestation, and may eventually cause death. RELATED SPECIES: *Arceuthobium* species, with yellowish, orangy, or brown branches, are leafless parasites on pines in the Trans-Pecos.

KEY TO MISTLETOES

1a. Leaves inconspicuous, reduced to scales; parasitic on junipers and Arizona cypress in Trans-Pecos mountains . . . .
. . . . *Phoradendron juniperinum*

1b. Leaves not reduced to scales . . . . go to 2

2a (1b). Parasitic on junipers in Trans-Pecos mountains and the Western Edwards Plateau; leaves narrow, less than 1 inch (2.5 cm) long and ³⁄₁₆ inch (5 mm) wide . . . .
. . . . *Phoradendron hawksworthii* (*P. bolleanum*)

2b. Parasitic on broadleaf trees; leaves broader, elliptic to circular, ½–2 inches (1.2–5 cm) long . . . . go to 3

3a (2b). Parasitic on oaks in Trans-Pecos mountains (very rarely on mesquite); plants grayish or yellowish-green, with minute hairs; flowers in summer; fruit hairy at tip (hand lens) . . . .
. . . . *Phoradendron coryae*

3b. Parasitic on hackberry, mesquite, ash, willow, sycamore, cottonwood, acacia, walnut, and other trees, occasionally on oaks; flowers from winter to spring; fruit not hairy . . . . go to 4

4a (3b). Near the Rio Grande from El Paso to Presidio, mainly on cottonwoods; at elevations below 3,500 feet; plant not hairy . . . .
. . . . . . *Phoradendron macrophyllum* subsp. *cockerellii* (*P. tomentosum*)

4b. Abundant throughout most of Texas at low elevations or to 5,700 feet; plant hairy or not (hand lens) . . . .
. . . . *Phoradendron tomentosum* (photo 382) (*P. serotinum*)

LICHENS

383. Lichens are bizarre organisms composed of a fungus and an alga or cyanobacterium growing together. The alga manufactures sugars for both organisms; the roles of the fungus—in anchoring the organism and trapping moisture—are not essential for the survival of the alga. Many species grow in Texas, some on bare rock, others on tree bark. Lichens can survive extremes of heat, cold, and aridity, often becoming desiccated in dry seasons. Important soil builders, they break up rock and wood chemically, by releasing weak acids into the substrate, and mechanically, by sending root-like threads into cracks. The threads shrink and swell when the temperature changes, causing the cracks to enlarge. Extremely slow-growing, patches of lichens found on rock surfaces may be well over a hundred years old. Do not scrape them off rocks. Lichens appear in a wide array of colors, some yielding brightly colored dyes for wool.

MOSSES

384. Mosses are green plants that form mats on moist soil, coat moist bases of tree trunks, or grow on bare rock. Mosses, which reproduce by spores, lack true leaves but have soft green leaflike structures for photosynthesis. They usually follow lichens as secondary soil builders, or they may be the primary components of the soil, forming a layer of organic material on which flowering plants can grow. Sphagnum moss forms large mats in eastern bogs. The mats eventually solidify into peat, which is used in the nursery trade and as a source of fuel. Some moss species tolerate desert conditions, becoming completely desiccated until moisture returns, but most grow in wet areas, along streams, seeps, and ponds and in bogs. ❧

Appendix A

How to Identify
Common Plant Families

One of the most valuable tools for plant identification is knowledge of the characteristics of a plant family. Here we focus on 17 of the families of trees, shrubs, and wildflowers that you are most likely to find in Texas and in many other countries of the world. As you read the characteristics, compare them with the photographs of plants from those families. With each family, we have listed some popular cultivated plants as well as wild members.

For many families, the characteristics of the genus are basically the same as those of the family, as in the genus *Acer*, in the maple family, Aceraceae. Refer to genus descriptions to assist you in learning about plant families.

Amaryllidaceae (Amaryllis Family)
Iridaceae (Iris Family)
Liliaceae (Lily Family)

Herbaceous plants, roots often bulbs, corms, or tuberous rhizomes; leaves alternate, also basal, usually slender, grasslike; flowers with radial symmetry or nearly so; usually with 6 colorful tepals in 2 rows (some species with 3 colorful petals, 3 green sepals).

Amaryllidaceae—6 stamens, rarely 3; ovary is below base of tepals, buried in tip of flower stem (an inferior ovary). EXAMPLES:

Daffodil, rain lily. NOTE: Some botanists consider the Amaryllidaceae and the Agavaceae to be part of the Liliaceae. Many species are poisonous if eaten.

Iridaceae—Inner 3 petals may differ from outer 3 in shape and size; 3 stamens; ovary inferior; basal leaves folding in half and partially enclosing adjacent leaf. EXAMPLES: Iris, blue-eyed grass. Many species are poisonous if eaten.

Liliaceae—6 stamens, rarely 3; ovary sits above base of tepals (a superior ovary). EXAMPLES: Lily, wild onion. Contains both edible and several highly toxic species.

Apiaceae (formerly Umbelliferae), Carrot Family

Herbaceous plants (rarely shrubs or trees); stems often hollow; leaves alternate, pinnately compound or highly dissected, petioles sheathing the stem; flowers minute, 5-petaled, in umbels or heads; fruit dry, 1-seeded (example: dill seed). EXAMPLES: Carrot, parsley, poison hemlock. Contains both edible and several highly toxic species.

Asclepiadaceae, Milkweed Family

Herbaceous plants or vines; sap usually milky; leaves simple, often opposite or whorled; unusual flowers with 5 petals (see *Asclepias* and *Matelea* species); flowers usually in umbels; fruit a pod, splitting open, filled with seeds usually attached to silky hairs. EXAMPLE: Milkweed. Many species toxic.

Asteraceae (formerly Compositae), Sunflower or Composite Family

Usually herbaceous plants (rarely trees, shrubs, or vines); sap milky or not; leaves simple or compound, often alternate or in basal rosettes; what may appear to be a single showy flower is actually a dense cluster (a head) of numerous tiny flowers; may have 1 or 2 types of flowers per head—ray flowers (petals united to form a single elongated, flattened strap) and/or disk flowers (petals form a tiny tube or funnel) (a third type of bilateral flower is rare in the family); head surrounded by numerous small leaflike bracts; fruit an achene, a 1-seeded, dry fruit (example; sunflower seed). EXAMPLES: Chrysanthemums, daisies, marigolds, asters. NOTE: This is the second largest family of flowering plants in the world. Contains many economically valuable flowers.

Brassicaceae (formerly Cruciferae), Mustard Family

Herbaceous plants; leaves usually alternate, often also in a basal rosette (basal leaves often pinnately lobed); 4 petals, not united, forming the shape of a cross (cruciform); 6 stamens, 2 usually short; fruit a silique, a many-seeded capsule characteristic of the family, quite variable in shape, either long and narrow or short and broad; after fruit splits open, only central papery segment remains. EXAMPLES: Broccoli, mustard, shepherd's purse, and numerous other edible species.

Cactaceae, Cactus Family

Succulents, stems swollen, fleshy, usually green; usually lacking leaves; areoles (pits or protrusions) scattered across surface of plant; spines, when present, in clusters at areole; flowers with radial symmetry, numerous petals and stamens; fruit a many-seeded berry, often fleshy. EXAMPLE: Prickly pear. Many rare species.

Fabaceae (formerly Leguminosae), Legume or Bean Family

Trees, shrubs, wildflowers, and vines; leaves alternate, usually compound (palmate or pinnate); margins of leaflets usually entire; flowers usually with 5 petals but in 3 general forms—(1) flowers papilionaceous (strongly bilateral, pea-type, with the top petal known as the banner, 2 side petals forming wings, and 2 bottom petals forming the keel (example: bluebonnets), (2) flowers only slightly bilateral, with petals almost equal in size and shape (example: sennas), (3) flowers with radial symmetry, but minute, many in a dense round head (resembling a puffball) or cylindrical spike (resembling a bottle brush) with stamens protruding and colorful (examples: huisache, mimosa tree, mesquite); fruit a bean pod, shape highly variable in different genera. NOTE: This is perhaps the third largest plant family in the world. Contains numerous economically important food plants and many species that are highly poisonous.

Lamiaceae (formerly Labiatae), Mint Family

Herbaceous plants or shrubs; stems square; leaves usually simple, opposite or whorled, often aromatic; flowers usually strongly bilateral, 5 petals united, forming 2 lips; 5 sepals, united, often bilateral; flowers either in tight cluster in the leaf axils or in spikes or whorls above leaves; 2 or 4 stamens; fruit a cluster of 4 tiny 1-seeded nutlets. EXAMPLES: Peppermint, catnip, horehound. NOTE: Many mints are useful as teas, however this family is similar in appearance to Scrophulariaceae and Verbenaceae, both of which have some highly toxic species.

Malvaceae, Mallow Family

Herbaceous plants, shrubs, or rarely trees; leaves simple, alternate, often palmately lobed or veined; flowers with radial symmetry, with 5 petals; many stamens, often forming a tubular column; fruit with 5 or more segments, dry or fleshy. EXAMPLES: Hollyhocks, hibiscus, Turk's cap. Many economically important species.

Orchidaceae, Orchid Family

Herbaceous wildflowers, growing in soil, tropical species often resting on trees (epiphytic, extracting water from the air); leaves simple, usually alternate, margins entire; flowers strongly bilateral, with 3 petals, 3 sepals (sometimes colorful, showy), 1 or 2 stamens. EXAMPLES: Vanilla, lady's slipper, ladies' tresses. NOTE: Though orchids are rare in the U.S., the Orchidaceae is the largest family of flowering plants in the world. Many economically important species.

Ranunculaceae, Buttercup Family

Mostly herbaceous plants; leaves palmately compound, petioles sheathing the stem; flowers with radial or bilateral symmetry, often with numerous tepals, stamens, and pistils. EXAMPLES: Buttercup, anemone. Many species toxic.

Rosaceae, Rose Family

Trees, shrubs, and wildflowers; leaves simple or compound, alternate; flowers with radial symmetry, 5 petals, 5 sepals; many stamens; 1 to many pistils; fruit dry or fleshy. EXAMPLES: Rose, cherry, plum and numerous other species with edible fruit. Seeds, leaves, and bark of some species contain cyanide-forming glycosides.

Scrophulariaceae, Figwort or Snapdragon Family

Mostly herbaceous plants (rarely vines, trees, or shrubs); leaves simple, alternate or opposite; flowers usually bilateral, 5 (or 4) united petals, often forming 2 lips; 4 (or 2) stamens, often with sterile 5th stamen (may be hairy); fruit a many-seeded capsule or berry. EXAMPLES: Snapdragon, penstemon. NOTE: This family is similar to Lamiaceae in appearance. Some species toxic.

Solanaceae, Nightshade Family

Herbaceous plants, vines, shrubs, or small trees; leaves simple; flowers usually with radial symmetry, 5 petals, united, often star-, bell-, or trumpet-shaped; 5 (4-6) sepals, not united; usually 5 stamens, often merging together to form a prominent beak; fruit a fleshy berry or dry capsule. EXAMPLES: Tomato, potato, silverleaf nightshade. Contains both edible and highly toxic species.

Verbenaceae, Verbena Family

Herbaceous plants, shrubs, or trees; stems often ridged or square; leaves usually opposite or whorled, often aromatic; flowers with 5 petals united to form short tube with lobes nearly flat; petals not equal in size or shape but usually not strongly bilateral; 5 sepals; 4 stamens (2 short); flowers in loose or dense spikes or round or flat-topped heads. EXAMPLES: Lantana, verbena. This family is similar in appearance to Lamiaceae. Most species either toxic or inedible. ❧

Appendix B

Plant Watching: A New Nature Hobby

The hundreds of thousands of birdwatchers in the United States someday may have to step aside for a new generation of nature lovers—plant watchers. Though plants don't sing, trees, shrubs, wildflowers, and other plants have many of the same qualities that make birds exciting, plus a few more. They are as colorful, sometimes as hard to find, and as challenging to identify. Plants fill us with wonder about their bizarre sexual strategies, chemical defenses, ingenious methods of seed dispersal, complex adaptations for surviving in a harsh world, and multitude of beneficial uses. Included in the plant kingdom are the largest and longest-living organisms on earth. By supplying food and oxygen, plants provide the foundation for all life. Through millions of years of coevolution, plants and animals have developed intriguing and mystifying relationships. The more we learn about plants, the more we are awed by their beauty, variety, and fascinating life histories.

The first step in plant watching is identification. Birdwatchers have it comparatively easy. Fewer avian species occur in all the United States than plant species in many state parks in Texas. Learning to recognize the main plant families helps identification greatly. If you cannot identify a plant by using a field guide, document the plant's characteristics in the field (see the sample documentation sheet on page 313). Positive identification may require consulting botanical manuals (and learning the specialized terminology used by botanists), using a hand lens or microscope, visiting an herbarium, or consulting a botanist. To the serious plant watcher, the complexity of a plant only adds to its mystery and interest.

Plant enthusiasts have two distinct advantages over birders and other wildlife watchers. As subjects of study and photography, plants stay in one place. Also, you can do more than take a photo and add the plant to your life list of species seen. When plants are numerous, you can collect a specimen and preserve it. A plant watcher can build a private herbarium and place duplicates in university collections. Like the records of the worldwide cadre of bird-watchers, the data compiled by an extensive network of plant enthusiasts can provide scientists with valuable information about distribution, abundance, and ecology.

GUIDELINES FOR COLLECTING PLANTS

---

1. Collect only when the plant is abundant and when you have the landowner's permission. Picking plants on any state or federal property is illegal without a special permit. Many native plants are rare and in danger of extinction and should never be collected.

2. Try to collect samples with flowers or fruit, preferably both. Unless you are collecting for scientific studies, leave the roots of herbaceous perennials undisturbed. Always leave an abundant supply of living plants to produce seed for next year's crop. On trees and shrubs, carefully prune a twig with leaves, flowers, or fruit.

3. Carry a plastic bag in the field to hold your specimens. Spray a tiny bit of water into the bag. This will help prevent wilting for an hour or so until you can press the plants.

4. As soon as possible after collecting a specimen, carefully spread it between sheets of newspaper. Arrange it to look natural, with some leaves bottom-side up. Protect delicate flower parts from sticking to the newspaper by placing them between absorbent paper towels. If you have a number of specimens, place the newspapers in layers between pieces of smooth-faced corrugated cardboard or blotter paper (to absorb moisture). Press the plants with heavy books or a plant press, and keep them in a warm place for 3 to 10 days. Many herbariums require specimens to be frozen at a low temperature or heated in a microwave to kill insects.

5. In a field notebook, record the scientific and common names, date, location (city, county, directions), flower color, plant type (herb, vine, shrub, tree, etc.) and height, population variations, soil type, exposure, abundance, and associated plants. Using a preprinted field documentation sheet simplifies identification and record keeping (see page 311 for a sample). Assign the plant a unique number, and write it on the drying sheet. By referring to accurate field notes from decades past, modern botanists have been able to locate populations of plants long thought extinct.

6. When the plant is thoroughly dried, remove it from the newspaper. If the dried specimen sticks to the paper, gently bend the paper downward until the specimen pops free. Use white glue to attach the plant to construction paper or a special 11½-by-16½-inch acid-free herbarium sheet. Dilute the glue with water. You can paint the glue on the specimen; or spread glue on a sheet of glass or hard plastic, set the specimen on the glass, and then lift if off. Place the glued specimen on the paper.

7. Place seeds and other loose parts in an envelope glued to the sheet. Reinforce large parts either by sewing them onto the paper with heavy thread or string or by cutting thin strips of paper or a special gummed cloth, laying them across the specimen, and glue the strips onto the herbarium sheet.

8. Record the field number, field information, and the collector's name on a 3-by-5-inch label glued to the lower right hand corner of the herbarium sheet. Without proper documentation, the pressed plant has little value to others who may refer to your collection. File the completed sheet in manila folders, and store them in a sealed container with moth crystals to kill insects. If properly preserved, the specimen should last for decades. Many herbariums contain sheets more than one hundred years old.

MAKING A PLANT PRESS

A small plant press of 1-by-¼-inch (or ⅜-inch) slats is easy to construct. For a 12-by-18-inch press, cut eight 18-inch-long pieces and eight 12-inch pieces. For each side, align four 18-inch strips parallel and approximately 2½ inches apart. Place four 12-inch strips across them, one at each end and two equally spaced apart in the middle. Glue and nail, or screw, the pieces together. To use the press, place

A plant press is easy to construct. Use this sketch to make your own.

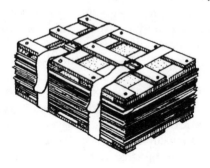

the newspaper and plants between sheets of cardboard or blotter paper, and tighten the press with two nylon straps.

A plant press is easy to construct and is an excellent tool for preserving plants. ❧

Herbarium Documentation Sheet

Date	
Scientific name	Family
Common name	Type of plant

Location

Habitat

Abundance (abundant, common, rare, scattered)

STEM (erect/spreading/twining)   Color/shape/texture

Height of plant

FLOWER Symmetry (radial, bilateral, papilionaceous)

Shape	Inflorescence type
Flower diameter	Inflorescence diameter
Length of pedicel	(terminal or axillary)

FLOWER PARTS	Color	Number	Attachment	Length	Width/Shape
Petals					
Ray flowers					
Disk flowers					
Sepals					
Bracts					
Stamens					
Pistil					
Stigma					
Style					
Ovary					

FRUIT (type, size, number of seeds)

LEAF	Shape	Margin	L	W	Texture (top/bottom)	Petiole	Arrangement (alt/opp/whorl)
Upper stem							
Lower stem							
Basal leaves							

Miscellany (distinctive odor, leaf venation, bark, etc.)

References

Ajilvsgi, Geyata. (1979). *Wildflowers of the Big Thicket, East Texas, and Western Louisiana*. College Station, Texas: Texas A&M University Press.

———. (1984). *Wildflowers of Texas*. Fredricksburg, Texas. Shearer Press.

———. (1991). *Butterfly Gardening for the South*. Dallas, Texas: Taylor Publishing.

Andrews, Jean. 1993. *The Texas Bluebonnet*. Austin, Texas: University of Texas Press.

Correll, Donovan Stewart, and Marshall Conring Johnston. 1970. *Manual of the Vascular Plants of Texas*. Renner, Texas: Texas Research Foundation.

Cox, Paul W., and Patty Leslie. (1988). *Texas Trees: A Friendly Guide* (third edition). San Antonio, Texas: Corona Publishing Co.

Damude, Noreen, Kelly Bender, Diana Foss, and Judit Gowen (1999). *Texas Wildscapes Gardening for Wildlife*. Austin, Texas: University of Texas Press.

Enquist, Marshall. (1987). *Wildflowers of the Texas Hill Country*. Austin, Texas: Lone Star Botanical.

Evans, Douglas B. (1998). *Cactuses of Big Bend National Park*. Austin, Texas: University of Texas Press.

Everett, James, and D. Lynn Drawe. (1992). *Trees, Shrubs, and Cacti of South Texas*. Lubbock, Texas: Texas Tech University Press.

Flora of Texas Consortium. http://www.csdl.tamu.edu/FLORA/ftc/ftcabout.htm

Gould, Frank W. (1998). *Common Texas Grasses: An Illustrated Guide*. College Station, Texas: Texas A&M University Press.

Hatch, Stephen L., and Jennifer Pluhar. (1993). *Texas Range Plants*. College Station, Texas: Texas A&M University Press.

Hatch, Stephen L., Joseph L. Schuster, & D. Lynn Drawe. (1999). *Grasses of the Texas Gulf Prairies and Marshes.* College Station, Texas: Texas A&M University Press.

Haukos, David A., and Loren M. Smith (1997). *Common Flora of the Playa Lakes.* Lubbock, Texas, Texas Tech University Press.

Hodoba, Theodore. (1995). *Growing Desert Plants: From Windowsill to Garden.* Santa Fe: Red Crane Books.

Irwin, Howard S., and Mary Motz Wills. (1961). *Roadside Flowers of Texas.* Austin, Texas: University of Texas Press. (Out of print).

Kirkpatrick, Zoe Merriman. (1992). *Wildflowers of the Western Plains: A Field Guide.* Austin, Texas: University of Texas Press.

Liggio, Joe, and Ann Orto Liggio. (1999). *Wild Orchids of Texas.* Austin, Texas: University of Texas Press.

Loughmiller, Campbell, and Lynn Loughmiller. (1984). *Texas Wildflowers.* Austin, Texas: University of Texas Press.

Mahler, William F. (1988). *Shinners' Manual of the North Central Texas Flora.* Dallas, Texas: Southern Methodist University Herbarium. (Out of print).

Metzler, Susan, and Van Metzler. (1991). *Texas Mushrooms: A Field Guide.* Austin, Texas: University of Texas Press.

Miller, George Oxford. (1991). *Landscaping with Native Plants of Texas and the Southwest.* St. Paul, Minnesota: Voyageur Press.

Muller, Cornelius H. (1950). *The Oaks of Texas.* Dallas, Texas: Texas Research Foundation. (Out of print).

Native Plant Society of Texas. http://www.npsot.org

Niehaus, Theodore F. 1998. *A Field Guide to Southwest and Texas Wildflowers.* Peterson Field Guide Series. New York: Houghton Mifflin Co.

Nixon, Elray S. (1985). *Trees, Shrubs, and Woody Vines of East Texas.* Nacogdoches, Texas: Bruce Lyndon Cunningham Productions. (Out of print).

Nokes, Jill. (2001). *How to Grow Native Plants of Texas and the Southwest.* Austin, Texas: University of Texas Press.

Powell, Michael, A. 1998. *Trees and Shrubs of the Trans-Pecos and Adjacent Areas.* Austin, Texas: University of Texas Press.

Richardson, Alfred. (1995). *Plants of the Rio Grande Delta.* Austin, Texas: University of Texas Press.

———. (2002). *Wildflowers and Other Plants of Texas Beaches and Islands.* Austin, Texas: University of Texas Press.

Rogers, Ken E. (2000). *The Magnificent Mesquite.* Austin, Texas: University of Texas Press.

Rose, Francis L., and Russell W. Strandtmann. (1986). *Wildflowers of the Llano Estacado.* Dallas, Texas: Texas Tech University Press.

Simpson, Benny J. 1988. *A Field Guide to Texas Trees.* Austin, Texas: Texas Monthly Press. (Although the book is out of print, the content of the book is now available on the Internet at http://dallas.tamu.edu/native/index.html)

Taylor, Richard B., Jimmy Rutledge, and Joe G. Herrera. (1997). *A Field Guide to Common South Texas Shrubs*. Austin, Texas: University of Texas Press.

Texas A&M University Bioinformatics Working Group. http://www.csdl. tamu.edu/FLORA/bwgprorb.htm (Extensive database on Texas plants and animals)

Tveten, John and Gloria Tveten. (1997). *Wildflowers of Houston and Southeast Texas*. Austin, Texas: University of Texas Press.

Tull, Delena. (1987). *Edible and Useful Plants of Texas and the Southwest: A Practical Guide*. Austin Texas: University of Texas Press.

USDA Natural Resources Conservation Service—Has an extensive plants database. http://plants.usda.gov/

University of Texas—Flora of Texas project. http://www.biosci.utexas. edu/prc/

Vines, Robert A. (1960). *Trees, Shrubs, and Woody Vines of the Southwest*. Austin, Texas: University of Texas Press. (Out of print).

———. (1977). *Trees of East Texas*. Austin, Texas: University of Texas Press. (Out of print).

———. (1982). *Trees of North Texas*. Austin, Texas: University of Texas Press. (Out of print).

———. (1984). *Trees of Central Texas*. Austin, Texas: University of Texas Press.

Warnock, Barton H. (1970). *Wildflowers of the Big Bend Country, Texas*. Alpine, Texas: Sul Ross University.

———. (1974). *Wildflowers of the Guadalupe Mountains and the Sand Dune Country, Texas*. Alpine, Texas: Sul Ross University.

———. (1977). *Wildflowers of the Davis Mountains and Marathon Basin, Texas*. Alpine, Texas: Sul Ross University.

Wasowski, Sally, and Andy Wasowski. (1997). *Native Texas Gardens: Maximum Beauty, Minimum Upkeep*. Houston, Texas: Gulf Publishing.

———. (2000). *Landscaping Revolution Garden with Mother Nature, Not Against Her*. New York: McGraw-Hill.

Weniger, Del. (1984). *Cacti of Texas and Neighboring States*. Austin, Texas: University of Texas Press.

West, Steve. (2000). *Northern Chihuahuan Desert Wildflowers*. Guilford, Conn.: Falcon Press.

Glossary

Asterisks designate terms that are illustrated in the flower and leaf glossaries at the end of the book.

achene—a small dry fruit containing a single seed (example, sunflower seed).

aggregate fruit—a dense cluster of berries (example: dewberry or blackberry.)

allopatric speciation—occurs when two species that normally live in different locations are brought together, hybridize, and produce viable offspring. Compare with **sympatric speciation**.

*alternate—refers to an arrangement of plant parts, such as leaves, that is neither opposite nor whorled.

annual—a plant that lives for only one growing season.

*anther—the tip of the stamen, holding the pollen.

areole—a small protrusion or depression, such as those scattered across the surface of cactus stems.

*axil-the juncture of a leaf or flower with a stalk.

berry—a fleshy fruit without a hard stone (example: tomato).

biennial—a plant that lives for two years, usually producing seed the second year.

*bilateral symmetry—an object, such as a flower, possesses this symmetry if it forms mirror-image halves when cut along only one plane (example: bluebonnet); compare with **radial symmetry**.

*bipinnately compound leaf—in this text, a leaf composed of two or more pinnately compound sections joined at their bases (example: mesquite).

*blade—the broad, flat portion of a leaf or petal.

bloom—a white waxy coating, such as that found on the surface of some fruits or twigs. Also, a flower or blossom.

*bracts—usually small and leaflike structures occurring below the petals and sepals of a flower. In some species, bracts are colorful and showy, like petals (example: poinsettia).

*calyx—the sepals, collectively.

capsule—a dry fruit that is composed of more than one ovary and that splits open at maturity.

*catkin—a flower spike, often pendent in a tree or shrub. The tiny flowers usually lack petals and may bear only male or only female parts.

clasping—refers to a leaf base that partially or entirely surrounds the stem.

*compound leaf—a leaf divided into two or more units, with each unit (leaflet or pinna), resembling a separate leaf (compare with **simple leaf**). A leaf bud, when present, will be found at the base of the compound leaf.

corolla—the petals of a flower, collectively.

deciduous—falling off. Said of trees that lose their leaves in winter.

dioecious—with male and female flowers occurring on separate plants.

*disk flowers—the tiny tubelike flowers found on the heads of members of the sunflower family, Asteraceae (example: inner flowers on a sunflower). Compare with **ray flowers**.

*dissected leaf—a leaf blade that is deeply cut into many small sections (example: yarrow leaf or carrot leaf).

drupe—a fleshy fruit with a single hard stone (example: cherry). The stone may contain several seeds.

endemic—growing only in a limited area.

*entire margin—a leaf edge that has no teeth or lobes.

evergreen—said of trees that maintain leaves year-round, never losing them all at one time. In Texas, not all evergreens are conifers, and not all conifers are evergreen.

falcate—sickle-shaped.

gall—a swelling on a plant, such as those resulting when an insect or mite lays eggs in a twig, leaf, or branch.

glaucous—covered with a whitish waxy coating (bloom).

glochid—a small barbed hairlike or bristlelike spine usually clustered at the base of longer spines; found only in the *Opuntia* genus of cactus.

*head—a cluster of flowers densely packed on a receptacle. The cluster may resemble a single flower (example; sunflower, daisy).

herbaceous—refers to a plant that dies back in winter (or in the drought of summer). Usually refers to nonwoody plants, though some herbaceous plants may have a woody base that does not die back.

inflorescence—a flower cluster.

*lanceolate—lance-shaped; that is, long and narrow, fatter near the base, and diminishing to a point at the tip.

leaflet—one of the units of a compound leaf.

lenticels—spots or short horizontal lines visible on smooth bark, that occur at openings or pores and are sites for gas exchange (example: cherry bark).

*linear—long and very narrow, of uniform width.

*lobe—a segment formed by a large indentation on the margin of a leaf or flower part.

*margin—the edge of a leaf or leaflet.

monoecious—with male and female flowers separate but occurring on the same plant.

*oblanceolate—lance-shaped; that is, long and narrow, with the fatter end near the tip (compare with **lanceolate**).

*oblong—several times longer than wide, with uniform width.

obovate—egg-shaped, widest near tip (compare with **ovate**).

*opposite—refers to an arrangement of plant parts, such as leaves, in which two occur at one point on the stalk (compare with **whorl** and **alternate**).

*oval—a broad ellipse, with both ends the same width.

*ovary—the base of the pistil (the female organ of a flower), containing the embryos that develop into seeds.

ovate—egg-shaped, widest near base (compare with **obovate**).

*palmately compound leaf—a compound leaf in which all the leaflets are united at a single point (compare with **pinnately compound leaf**).

*palmately lobed—having large indentations that diverge from a single point (compare with **pinnately lobed**).

*panicle—a type of flower cluster in which the flowers, each on a short stem, are arranged along stalks that branch off a central stalk.

*papilionaceous flower—a particular pea-type flower of the legume family (Fabaceae) with bilateral symmetry (example: bluebonnet). The top petal is called the banner, the two side petals are the wings, and the two bottom petals form a tiny boatlike keel.

*pedicel—the stem of a single flower.

perennial—a plant that lives for more than one or two years.

*petals—the leaflike parts of a flower that are usually brightly colored and enclose the reproductive organs (compare with **sepals** and **tepals**). Collectively known as the corolla.

*petiole—the stem of a leaf.

*pinna (plural: **pinnae**)—the main division of a pinnately compound leaf.

*pinnately compound leaf—a compound leaf in which numerous leaflets are attached to each side of a central stem. The leaflets may be alternate or opposite on the stem (compare with **palmately compound leaf**).—***twice pinnately compound leaf**—a compound leaf in which the primary units (pinnae) are again divided into pinnately compound sections.

*pinnately lobed—having large indentations along two sides of a central axis (compare with **palmately lobed**).

*pistil—the female part of a flower, which includes the ovary, style, and stigma.

*raceme—a type of flower cluster in which the flowers, each on a short stem, are arranged along a central stalk.

*radial symmetry—said of an object, such as a flower or pie, that forms mirror-image halves when cut along more than one plane (compare with **bilateral symmetry**).

*ray flowers—flowers with petals united to form a single elongated, flattened strap; found on the heads of members of the sunflower family, Asteraceae (example: outer flowers on a sunflower). Compare with **disk flowers**.

rosette—a circular cluster of leaves or flowers.—***basal rosette**—a rosette at ground level.

samara—a dry winged fruit that does not split open at maturity (example: fruits of maple or ash).

*sepals—leaflike and usually green structures arranged in a whorl below the petals of a flower. Collectively known as the calyx. Some sepals are colorful and indistinguishable from petals (compare with **tepals** and **bracts**).

sessile—attached without a stalk.

sheathing—refers to a leaf base that forms a tube completely enclosing the stem.

silicle—a shortened **silique**, often as broad as long.

silique—the pod of plants of the mustard family, often long and narrow, consisting of two valves, two sutures, and a partition, with the seeds attached to the edge of the partition.

*****simple leaf**—a leaf with a single blade (compare with **compound leaf**). A leaf bud, when present, will be found at the base of the leaf.

*****spadix**—a spike on a fleshy receptacle, surrounded by a **spathe**.

*****spathe**—a bract that encloses a flower or flower cluster (see **spadix**).

*****spike**—a type of flower cluster in which the individual flowers lack stems and are attached directly to the long central stalk.

*****stamen**—the male part of a flower, usually composed of a long, slender stem (the filament) and an enlarged tip that holds the pollen (the anther).

stellate—star-shaped.

*****stigma**—the tip of the pistil (the female organ of a flower).

stipules—a pair of leaflike bracts at the base of the leaves of some plants.

*****style**—the stem of the pistil (the female organ of a flower).

succulent—juicy and fleshy.

sympatric speciation—occurs when two species that live in the same location hybridize and produce viable offspring. Compare with **allopatric speciation**.

*****teeth**—small notches on the leaf margin.

tepals—refers to both petals and sepals when they are indistinguishable from each other.

*****trifoliate**—having three leaves or leaflets.

*****umbel**—a type of flower cluster in which the individual flowers occur on stems of nearly equal length and the stems unite at the top of the flower stalk.

*****whorl**—more than two plant parts, such as leaves or flowers, attached in a circle around one point on a stalk.

Index

* = previously used scientific name. Boldface numbers refer to page numbers for photographs.

FLOWER GLOSSARY

Flower cross-section

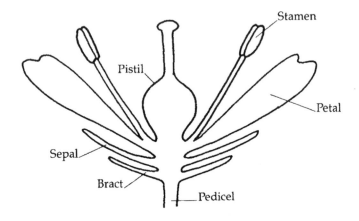

Radial symmetry   Bilateral symmetry

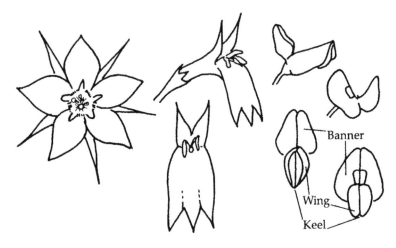

**Papilionaceous or
pea-type flower**

Inflorescences (flower clusters)

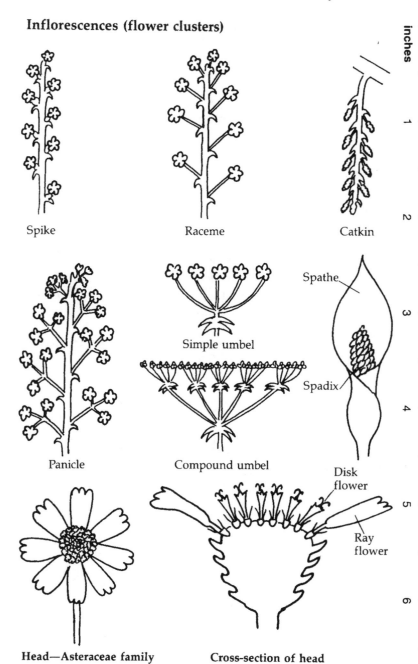

inches

Spike

Raceme

Catkin

Simple umbel

Spathe

Spadix

Panicle

Compound umbel

Disk flower

Ray flower

Head—Asteraceae family

Cross-section of head

LEAF GLOSSARY

Leaf parts

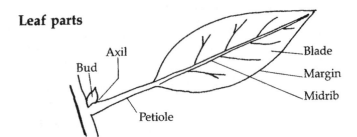

Axil

Bud

Blade

Margin

Midrib

Petiole

Leaf margins

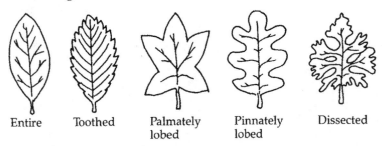

| Entire | Toothed | Palmately lobed | Pinnately lobed | Dissected |

Leaf shapes

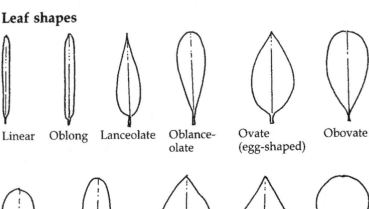

Linear Oblong Lanceolate Oblance-olate Ovate (egg-shaped) Obovate

Oval Elliptic Cordate (heart-shaped) Deltoid (triangular) Orbicular (round)

centimeters

Leaf types

Simple

Trifoliate

Palmately
compound

Bipinnately
compound

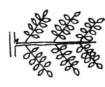
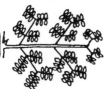

Pinnately
compound

Twice pinnately
compound

Thrice pinnately
compound

Leaf arrangement

Basal
rosette

Simple,
alternate

Simple,
opposite

Simple,
whorled

Leaves pinnately compound
and alternate

Leaves pinnately compound
and opposite